D1221434
CINCINNATI STAT
WITHDRAWN

THE SOLAR ENERGY ALMANAC

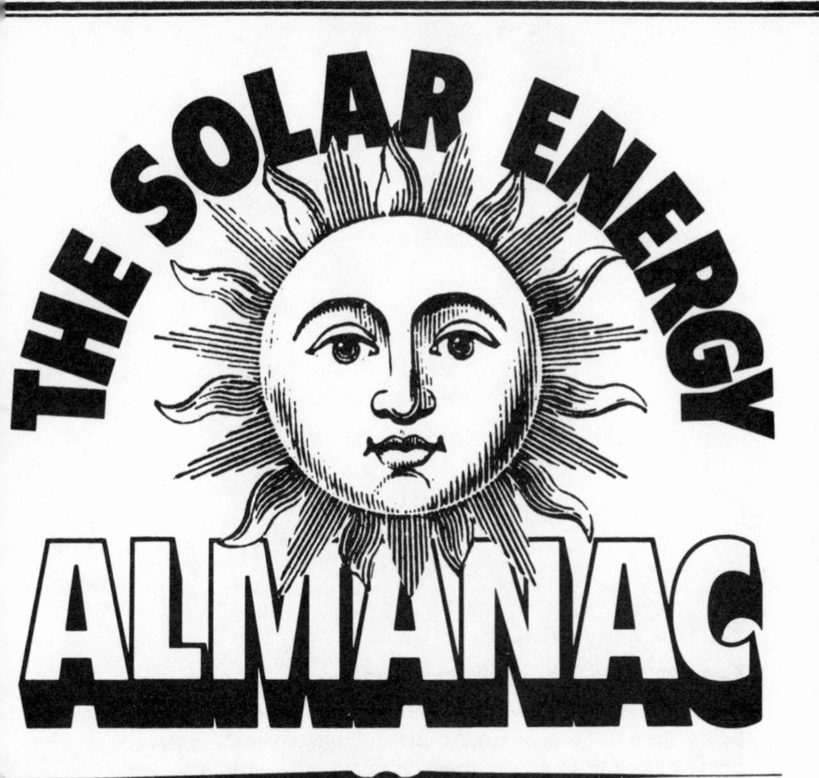

THE SOLAR ENERGY ALMANAC

EDITED BY MARTIN McPHILLIPS

Facts On File Publications
460 Park Avenue South
New York, N.Y. 10016

The Solar Energy Almanac

Published by Facts on File, Inc.
460 Park Avenue South, New York, N.Y. 10016

Library of Congress Cataloging in Publication Data:

Main entry under title:

The Solar energy almanac.

1. Solar energy—Handbooks, manuals, etc.
I. McPhillips, Martin.
TJ810.S6154 1982 621.47 82-9362
ISBN: 0-87196-727-8

10 9 8 7 6 5 4 3 2

Printed in the United States of America

A Sun Words book

Acknowledgements

This almanac would never have been possible without the help of work done under the solar program at the federal Department of Housing and Urban Development. That program laboriously sifted the sands of both passive and active solar heating technology and produced many of the documents on which much of the information in this book is based. The editor of this volume was privileged to serve as head writer and senior editor on some of the program's final and most conclusive publications.

The vast majority of illustrations contained herein have been borrowed from the wealth of HUD material. Therefore, unless other credit is given, it can be assumed that the illustrations were generated either by HUD or by the Department of Energy, the co-sponsor of the solar demonstration program.

Contributing editors

Sandra Oddo
Brad Shepp
Steve Hastie
Cole Whiteman
Ann Nocenti

Editorial assistant

Edye Weissler

Table of Contents

Introduction

What is solar energy? It seems that the question has several answers.

The term "solar energy" refers to simple sunlight; it also refers to the many ways that sunlight can be used to do things. This almanac is a basic book about the use of solar energy by individuals, communities, and nations.

Solar energy is a big subject. It is very diverse, and most of it is new to most people. There are too many sides to it to derive any benefit from force-feeding it to a reader. Therefore, this is a book that you can approach sideways—start anywhere, gravitate to the sections that interest you most.

If you are taken by the idea of personal independence, then at some point you will want to know about a most remarkable survival tool—*The Solar Greenhouse*, page 14. But the solar greenhouse belongs to a larger group of solar heating options that you should also know about—*The Passive Solar House*, page 2. Once you know about the options, you can check out how they are applied and combined—*Solar Houses Around the Country*, page 173.

Passive solar heating for the home is only the tip of the solar iceberg. There are exciting things on the horizon, like *Solar Satellites*, page 132, and remarkable *Solar Cells*, page 129.

And these are just a few of the ways in which solar can be used. Solar energy is a very broad topic, but people really owe it to themselves to be up on it. That information can enhance one's personal life. It is an alternative to traditional and/or ominous energy sources, a formidable survival tool, and a fascinating and consuming realm of knowledge.

Back now to what is currently the most appealing and significant use of solar—passive solar heating for houses. There was once a debate that argued for passive heating versus, well, *active* solar heating. It turns out that active solar is what most people think of when they think about solar—things like huge banks of solar panels mounted on the roof and large storage containers housed in the basement. But active solar has lost the debate. The huge expense and trouble of buying, engineering, installing, and maintaining an active solar space heating system are simply not worth it. (Active systems for water heating are not included in this denouncement. These are much simpler, more affordable, and are safe investments.) For space heating, passive solar makes so much more sense than active solar that there is no longer any basis for a debate. Cost comparison alone settles the issue: the upper price range for passive is about where the lower range for active begins.

Curiously enough, the federal income tax credit that allows for the purchase of solar heating equipment (40 percent of the cost, up to a maximum credit of $4,000) is far more easily taken for active systems. The IRS has made it difficult to get the credit for passive systems. The prejudice against passive solar is a residue from the early days of the Federal Solar Demonstration Program, when passive was viewed with suspicion because it went so much against the grain of traditional thinking. Even without the tax credit, passive is by far the better choice.

For solar energy as a whole, with regard to its wider significance, let me repeat some of the old points even though they may be tired: solar is inexhaustable; it is non-polluting; it is the very essence of energy independence both at the individual and at the national level; it is wholesome; and it will some day be seen as one of the great triumphs of one of our greatest assets: good old Yankee ingenuity.

The goal of this book is to present the clearest and most concise picture of solar energy that is possible. It is not a how-to-do it book in the sense that it will teach you the skills required to build a solar house or install a solar water heater. I don't expect the average American householder to learn carpentry, home construction, or plumbing simply to do something with solar. If you already have those skills, then this book gives you enough information to use them properly; if you don't have them, there are ample provisions here for helping you to find someone who does.

THE SOLAR ENERGY ALMANAC

1.
Sun Facts

The sun is actually closest to the earth in the dead of our winter, in January, at a distance of 91 million miles. It is farthest away in July, at 94 million miles. Our seasons are not determined by distance from the sun but by the tilt of the earth's axis in relation to the sun.

The sun is a fusion reactor that converts hydrogen to helium at its core at a temperature of approximately 25 million °F. The temperature at the sun's surface is a mere 10,000°F.

Of the sun's total radiant output the earth intercepts only about one part in 2 billion. But that's close to 35,000 times the amount of energy used by all of the inhabitants of earth in one year.

The diameter of the sun is about 865,000 miles, or approximately 110 times the earth's diameter. The sun's mass is 2.2×10^{27} tons, or about 335,000 times the earth's mass. The gravitional force of the sun is 28 times the earth's, so that a man who weighs 150 pounds on earth would weigh 2 tons on the sun. Wood, oil, natural gas, coal, and food are all fuels that, in reality, are forms of stored sunlight.

The sun is some 5 billion years old, and is burning at the rate of 4 million tons per second. It is expected to do this for another 6 billion years.

2. The Passive Solar House

Solar heating for the house is best accomplished through a method by which elements of *the house itself* collect sunlight and then store the resulting heat. The sunlight is collected through large south-facing windows or glazed walls. Its heat is absorbed and stored by specially constructed floors, walls, and/or room dividers made of concrete, brick, or block. Water-filled plastic tubes or metal drums can also be used to absorb and store the heat.

This method is known as **passive solar heating**. It is said to be *passive* because it involves few moving parts and does not rely extensively on another source of energy (i.e. electricity) for its operation.

The first modern passive solar houses (the ancient Greeks and Romans had known and used the concept) were designed and built in the early 1940s by George Keck, a Chicago architect. But the idea never really caught on until the 1970s, when people began to seek personal independence from high-priced conventional fuels.

Direct Gain

The method used by Keck, still considered the simplest and most common approach, is called **direct gain** (see *Figure 1*).

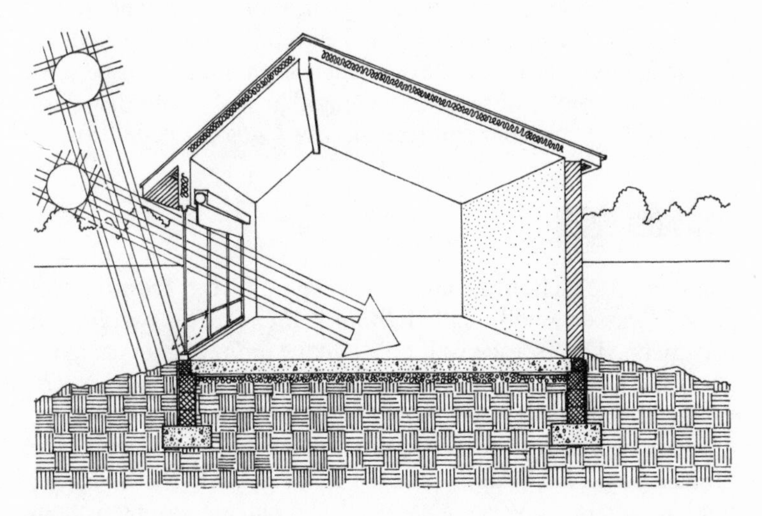

Figure 1. Direct gain

Sunlight passes through the south-facing glass directly into the living space. The sunlight strikes the walls and floors, which are concrete (or brick, or block), 4 to 8 inches thick, and sometimes even thicker. The surfaces of the walls and floors are painted or stained a dark color so that they will absorb the sunlight, changing it to heat. The heat will then be stored in the concrete mass. At night the heat will radiate back into the room as the indoor temperature begins to cool (heat radiates naturally from a warmer surface to a cooler one).

Not all of the solar heat is absorbed by the storage elements (called **thermal mass** in the jargon of solar energy). Some of it will heat the room air, keeping the indoors warm throughout the day. The walls and floors, in addition to holding heat for use at night, serve to keep the indoors from overheating during the day. Once again, water-filled containers—with proper structural support—will also serve this function.

It is essential to inhibit the loss of heat back through the large glass area at night. This problem is best solved with **movable insulation** used to cover the glass throughout the night, from sunset to sunrise. If properly designed, movable insulation can also be used during the summer to keep sunlight and heat out of the house during the day. Summer heat gain can also be prevented by a roof overhang or awning that will shade the glass from the high summer sun (see *Passive Cooling*, page 35).

Trombe Wall

Another approach, quite unlike direct gain, is the **Trombe wall** (which was originally called *Trombe-Michel* after two French scientists). The Trombe wall falls into the *indirect gain* category because sunlight is not collected directly by the living space.

The Trombe wall (*see Figure 2*) consists of a concrete, brick, or block wall 8 to 16 inches thick, built as the south-facing side of the house. This wall is covered on the outside with a single, double, or triple layer of glass or plastic glazing mounted about 4 inches in front of the wall's surface.

Solar energy passes through the glass, strikes the wall, and is absorbed and stored.

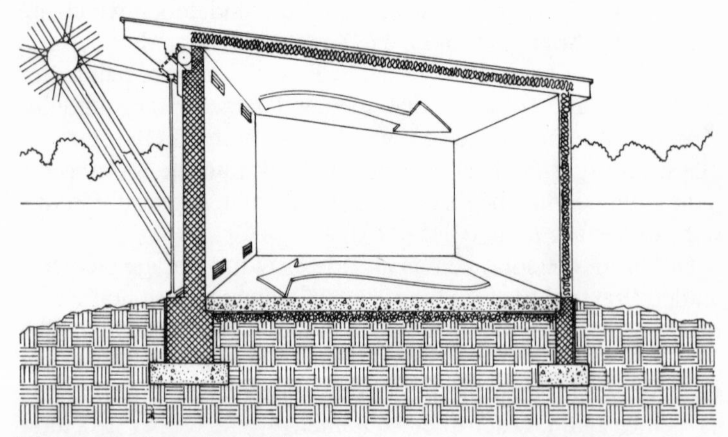

Figure 2. Trombe wall

The Trombe wall provides heat to the house in two ways. Over a period of several hours heat migrates through the wall and finally begins to radiate into the living space, usually late in the day and after the sun has set. For immediate heating, most Trombe walls also have two sets of vents—at the top and at the bottom of the walls—so that cool, floor-level room air is drawn into the space between the wall's surface and the glass. The air is heated, rises naturally, and circulates back into the room through the upper vents. This flow continues all the while the sun shines.

To counteract heat loss from the Trombe wall there should be an insulating curtain that can be drawn at night between the surface of the wall and the glass. The wall vents should be equipped with **backdraft dampers** to prevent warm room air from circulating back into the 4-inch space at night. Such a reverse flow would cool the room rather than warm it.

Water Wall

An important variation on the Trombe wall is the **Water wall** (see *Figure 3*). Here the masonry wall is replaced by water containers that stand between the windows and the room. It is an

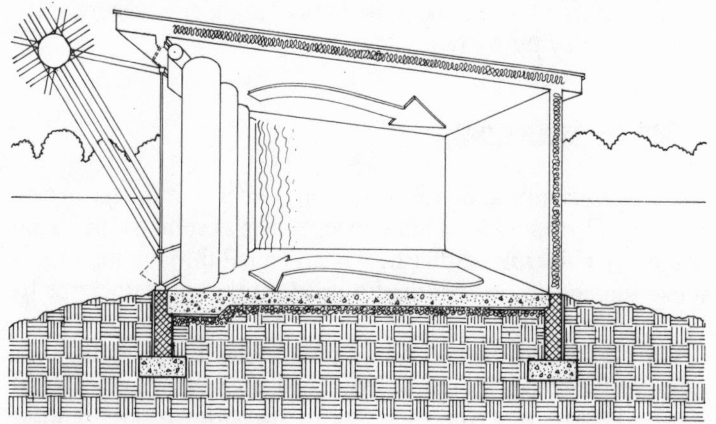

Figure 3. Water wall

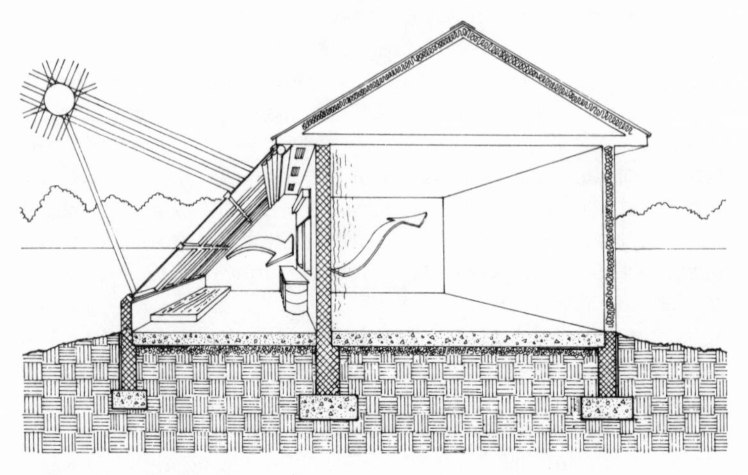

Figure 4. Solar greenhouse

increasingly popular approach. Water has a greater heat storing capacity than concrete, but it is also heavier and is not capable of serving as a structural support for the building.

Figure 3 shows a water wall that consists of tall, hard plastic tubes. Other durable containers will also work including 55-gallon drums, paint cans, and glass jars. As with the Trombe wall, heat loss through the glass area at night can be controlled with an insulating curtain.

Solar Greenhouse

The **solar greenhouse** (discussed at greater length starting on page 14) also includes such concepts as the **solar room, sunspace** and **solarium**. Solar heat is collected through the greenhouse glazing to be absorbed by water-filled containers or by masonry (see *Figure 4*).

The options available to solar greenhouse design make it the most versatile and perhaps the most desirable passive solar approach. *Figure 4* shows a few of the storage possibilities, including a masonry wall separating the greenhouse from the

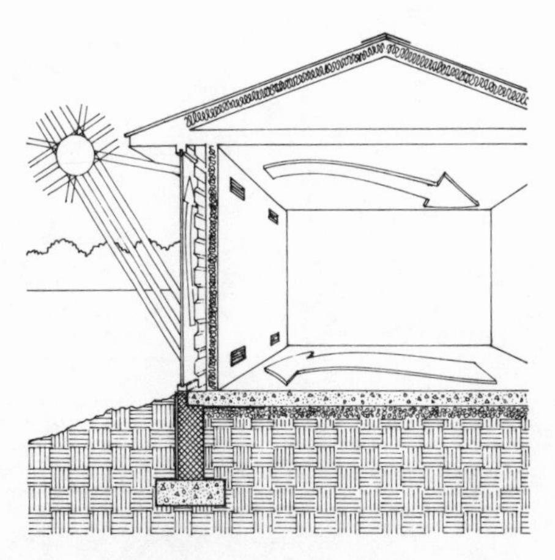

Figure 5. Thermosiphoning collector built into south wall

main living space, 55-gallon drums filled with water inside the greenhouse, and potting beds. A water wall could be used in place of the masonry wall, and a rock storage bed can be added either below the greenhouse or in the adjoining basement.

Thermosiphon Collector

Thermosiphon literally means *heat siphon*. It describes the natural tendency of air (or water) to rise as it gets warmer, in this case inside a collector panel attached to a house. This is the same way air circulates and is warmed as it passes upward through the space between a Trombe wall's surface and the glass that covers it.

This same process can be used in a collector that employs a sheet of corrugated metal with its surface painted black (*see Figure 5*). The aluminum is covered with glass or plastic glazing, leaving an inch or two of air space.

A thermosiphon collector can be built either into a south-facing wall (*as in Figure 5*) or at a lower level than the house (*Figure 6*). When it is part of a south wall, a thermosiphon panel

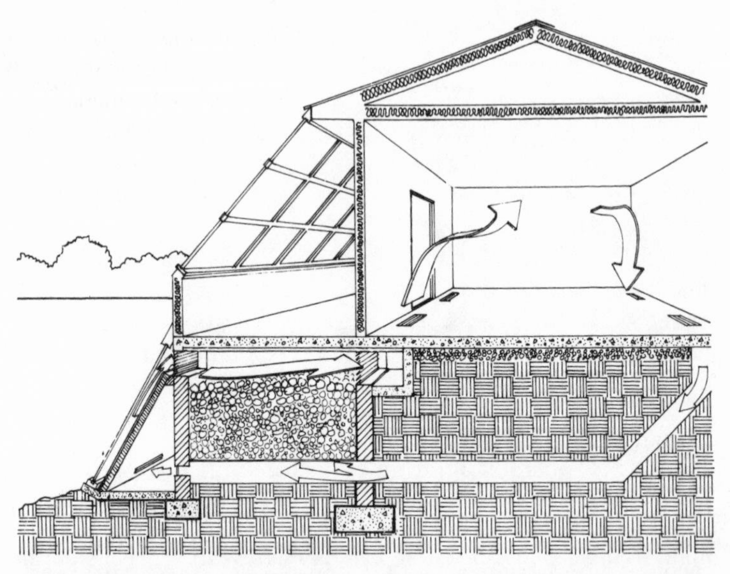

Figure 6. Thermosiphoning collector placed below a floor-level greenhouse and connected with a rock heat-storage bin

will deliver heat directly to the living space in the same way a Trombe wall will, but without the capacity to store any heat. When the collector is placed below the house it can be linked to a rock storage bin.

Sun-Tempering

Sun-tempering refers to capturing solar heat without providing for storage. An example is the use of a thermosiphon collector where none of the usual storage methods are employed. More typically, sun-tempering refers to a direct gain type of situation without the thick storage walls or floors. This, of course, lowers the cost but is generally thought to be less effective because it can overheat the living space during the day and stores no heat for use at night.

Sun-tempering has been used in a breed of house design known as **super-insulated**, where the building is so energy conserving that little heat is required. But as a general approach to solar heating, sun-tempering is judged here to be inadequate and superficial. New home buyers, looking to purchase a passive solar home, are therefore cautioned to make sure that they are getting a complete passive system and not simple sun-tempering.

Sun-tempering may be suited to buildings that are used primarily during the day, such as schools, shops, offices, and warehouses. But even in these cases the question of overheating because of the lack of storage should be addressed.

Combined Systems

It should be made clear that most passive approaches are not mutually exclusive. They can be used together. A Trombe wall can have large openings to allow direct gain. Direct gain can be designed to include a partial water wall. A solar greenhouse can, as shown, use direct gain, Trombe walls, and water walls.

The advantage to combining systems is that it lets the passive "system" serve the house, rather than the other way around.

See *Solar Houses Around the Country*, page 173, for examples of new house designs that feature one or more passive solar heating approaches. Also see *Passive Solar Construction Details*, page 57, for annotated drawings of various passive systems and elements. The details alone are not likely to be sufficient to allow you to design a solar house, but they provide basic information on points where solar houses may differ from conventional houses. See *Finding A Solar Designer/Builder*, page 12, for a brief summary of finding and evaluating a solar building professional.

3.
Heat Moves Naturally

Heat moves naturally in the direction of cold in accordance with the Second Law of Thermodynamics. There are three ways that it does so: conduction, convection, and radiation.

Conduction is heat moving through a solid, exactly the way it moves up the pot handle in the illustration, or through the wall of your house.

Convection is heat moving in air or liquid. Hot air rises—you can see its effect in a hot air balloon—because it is lighter and more buoyant than the air around it. Hot water will do the same thing.

Radiation is heat as an electromagnetic wave, unconnected to a solid, liquid, or air medium. Heat waves travel through your living space just as they do through outer space: they only demonstrate warmth when they strike a surface, like the earth, a wall, or your skin.

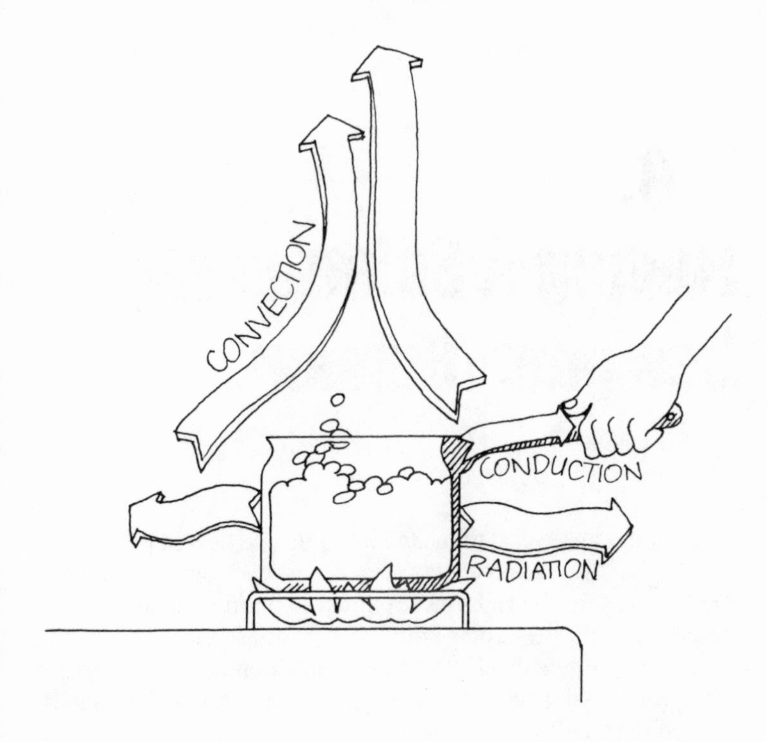

The three ways heat moves: radiation, conduction, and convection

4.
Finding A Solar Designer/Builder

There are now several thousand building and design professionals across the country, who have passive solar experience. Getting hold of the right professional to build your solar greenhouses or your new solar house is very important.

The investment of a little time in the search is likely to pay off. The project will proceed smoothly, and the work will be quality and on the mark.

Some methods for getting what you need:

1. The best way to find a solar professional is to get a referral from someone who has already used that professional and is very satisfied with the results. It is worth the trouble to start out with a few such referrals if they are available.

2. Ask the solar professionals for other references and check them, and if possible visit one of their completed projects.

3. Take a look at their portfolios of completed solar projects.

4. Discuss the design and construction procedures and check them with the passive solar construction details found on page 57 of this book. If you are having a solar greenhouse built, also refer to *The Solar Greenhouse*, page 14.

5. Specify in detail, as part of the contract, the work to be

done, the approximate dates for beginning and completion, and the return of all building components to normal condition after construction.

6. If you are unable to locate a solar professional through a personal or business recommendation, check the listing for your state in the *Sources of Information* section of this book, for an agency or organization that could provide you with referrals.

5.
The Solar Greenhouse

Figure 1. A typical solar greenhouse

on to an older one.

The key to the solar greenhouse's appeal is its remarkable versatility. It can be designed, built, and used to suit wide variations in taste, need, and budget. For instance, while it supplies heat, the solar greenhouse can be used both to start seedlings and grow vegetables. Or it can be a pleasant living space or sunspace or solarium where children can play and adults can sun themselves even though the temperature outdoors might be below freezing.

The primary concern of this article is with a greenhouse that can be built onto an existing home and used as a home heating system. A solar greenhouse built as part of a new house might be similar in design to this one, but it would be part and parcel of the larger building project and would not be approached as an addition. (*Solar Houses Around the Country*, page 173, has many examples of new houses built with solar greenhouses).

It Works Like This

First of all, a solar greenhouse should face south, or within 30 degrees thereof. During the winter the windows admit sunlight that strikes the darkened surfaces of 1) a concrete floor, 2) a brick wall, plus 3) water-filled drums, or 4) other specially designed heat-storage mass.

These surfaces convert the sunlight to heat. Some of this heat is absorbed into the concrete, brick, and water, where it will remain stored until the indoor temperature begins to cool after the sun sets. The heat not absorbed by the storage elements can raise the daytime air temperature inside the greenhouse into the 90° to 100°F range. As long as the sun shines, this heat can be circulated into the house, pulled by natural convection (heat always flows in the direction of cool) or pushed by a low horsepower fan.

As the sun sets and no longer provides warmth, the greenhouse windows should be covered with movable insulation (see

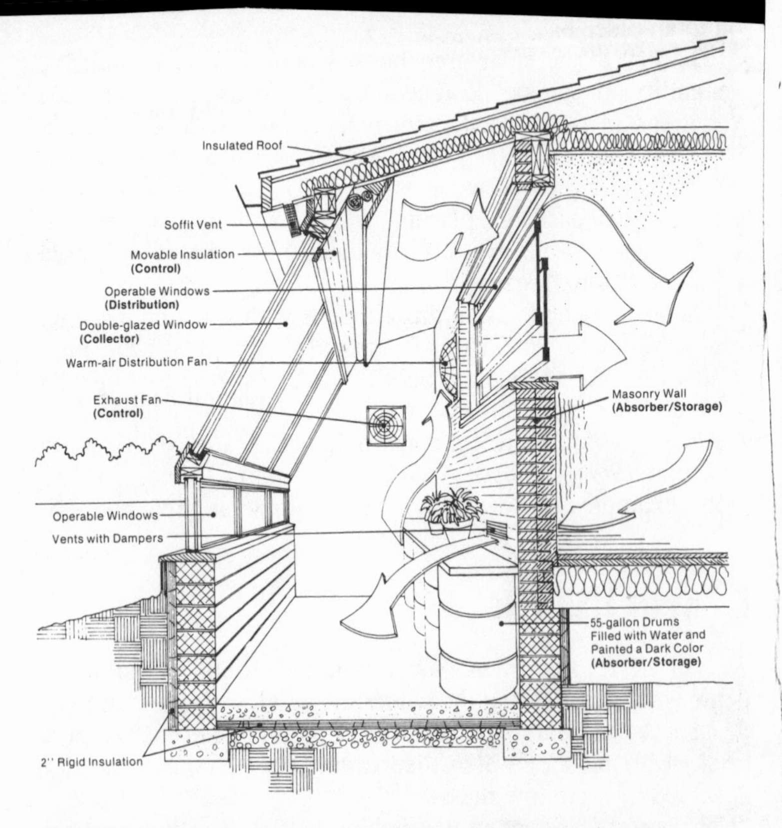

Figure 2. Cross-section of solar greenhouse showing construction details

Movable Insulation, page 40). This controls heat loss through the glass. The greenhouse will then remain warm as long as it can draw on the heat in its storage elements, and the house will continue to be warmed by the common brick storage wall it shares with the greenhouse. The storage elements will continue to supply heat until they are the same temperature as the air in the living space. The greenhouse is also a buffer zone that protects the main house from the outside cold.

The Passive Elements

Any passive solar heating arrangement should include five different but closely interconnected elements.

- A collector, i.e., a double-layer greenhouse window made of glass or plastic (clear or translucent).
- An absorber, i.e. the darkened (preferably black) *surfaces* of walls, floors, and water-filled containers inside the greenhouse.
- A storage mass, i.e. the concrete, brick, and/or water, which will retain the heat after it has been absorbed.
- A distribution system, i.e. pathways for natural heat flow or fans and ducts to get the heat into and around the house.
- A control system (or heat regulation devices), i.e. movable insulation to prevent heat loss through the glass at night; thermostats to activate fans.

It is essential that these five elements work together. For this reason all but the truly experienced do-it-yourselfer should, at the very least, consult with an *experienced* solar builder or designer to review plans and advise on construction procedures.

Some Options for the Attachment

The way that a solar greenhouse is attached to an existing house should reflect both the structure of the main house and the intended use of the greenhouse. There are at least four commonly used methods for making the connection.

- Separate the greenhouse from the house by a brick, block, or concrete wall (uninsulated). Solar heat absorbed and stored by the wall will migrate through the wall over several hours. This heat will reach the main living space late in the day and after the sun has set.
- Use glass doors and/or patio-door-sized windows to accomplish the separation. Inside the greenhouse, place tall water-

filled tubes or 55-gallon drums of water behind the stationary panes. If the greenhouse is not too deep, winter sunlight can pass first through the greenhouse and then through the glass windows and doors either to hit the water tubes or to enter the main living space directly. The tubes absorb and store some of the heat, but the room also receives immediate warmth.

- As above, the greenhouse is separated from the house by oversized windows and glass doors. But instead of water tubes, the sunlight strikes the masonry floor of the living space. The floor should be at least 4 inches thick—typically an uncarpeted concrete slab covered with ceramic tile or brick.
- Use *no* form of separation between the greenhouse and the main house during the day. Sunlight and heat will pass directly and easily from the greenhouse to the house. At night, however, movable insulation will be required either to cover the greenhouse glazing or to seal off the main structure from the greenhouse.

The best approach may be to combine options so that both structural and design requirements are satisfied.

Additional Considerations

By now it is fairly obvious that the design of a solar greenhouse requires attention to details. For instance, the design featured in *Figure 1* has an insulated roof. If the roof were glass, as it would be in most conventional *non*-solar greenhouses, the heat lost through the roof would be considerably greater since heat rises and accumulates near the roof. While glass is an adequate barrier against the loss of heat by radiation, it provides little resistance to the loss of heat by conduction.

When the greenhouse is used for growing plants or vegetables, this must balance with the heating function. Plants compete for light, leaving less energy available for the heat-absorbing surfaces. If too much light is absorbed for use as heat, plants will not cooperate and grow.

Movable night insulation is also a very important consideration. After the sun has set and the outdoor temperature drops,

Options for separating the greenhouse from the main house

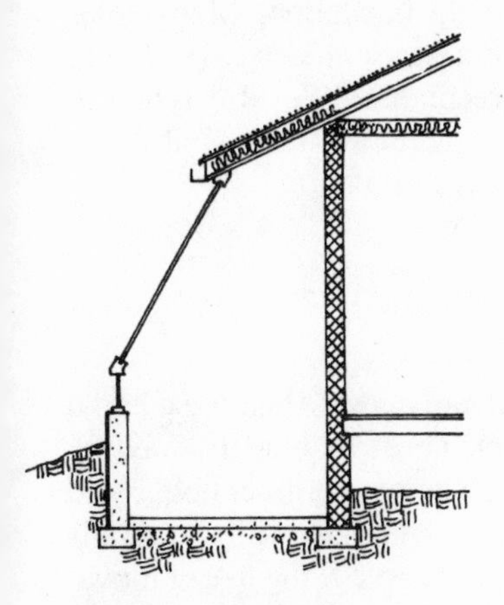

Figure 3. Masonry wall

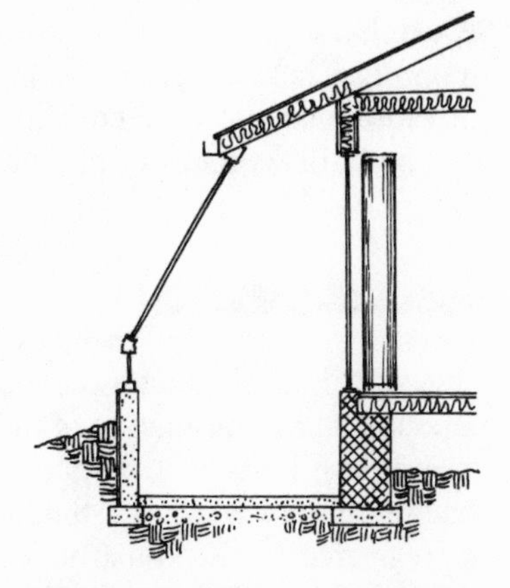

Figure 4. Glass, with water tubes

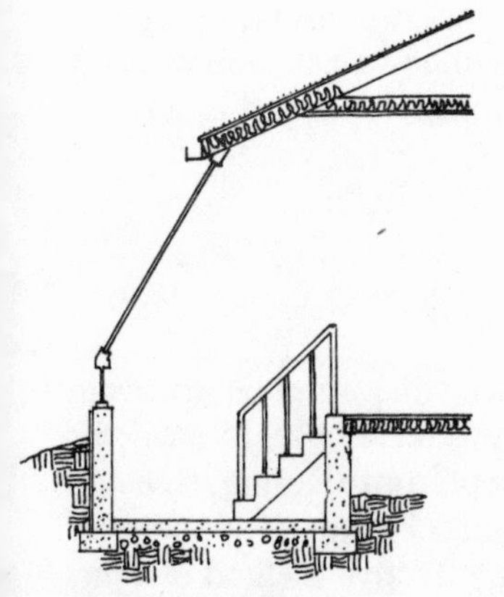

Figure 5. Open

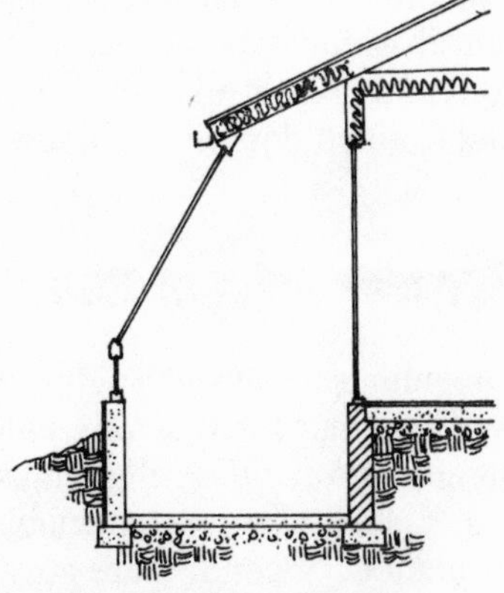

Figure 6. Glass alone

the large solar collection area of the greenhouse will lose heat at a steady rate and, of course, there will be no compensating sunlight. It is important to stop this heat loss.

The best approach is movable night insulation. Many solar researchers agree that night insulation is essential to deriving the maximum benefit from a solar greenhouse. Several types of movable insulation are now commercially available, but it can also be built with easily purchased materials.

Heat Contribution

The contribution a solar greenhouse can make to heating a home depends on a combination of factors. These include the way the greenhouse is used, the square footage of the greenhouse and the greenhouse windows, the local climate, how well the details of solar greenhouse construction are observed, the use of movable night insulation, and how heat is drawn into the house from the greenhouse.

A large greenhouse in a sunny climate, with a south-facing window area equal to one-fourth of the floor area of the house to which it is attached (e.g. 500 square feet of window for a 2,000 square foot house), might in theory supply as much as 75 percent of the heat used by that house. This would most likely occur in a house that is exceptionally energy conserving. Such a house would typically have storm windows and doors, extra wall and attic insulation, weatherstripped doors and windows, and caulked door and window frames.

Summer Adjustments

Greenhouses must be adjusted during the summer to avoid overheating. First of all, unless the greenhouse is used for summer growing, sunlight must be kept from entering, by using either stationary or retractable awnings, shades, or movable insulation. Second, there must be an effective method of venting any heat that does build up inside the greenhouse. (Summer greenhouse gardeners may have to find ways to get *more* light into the greenhouse, and provide for massive venting to prevent overheating.) The design shown in this article uses an exhaust fan like those used in kitchens for venting heat, but many solar greenhouses have roof vents that can be opened to allow heat to escape by natural convection. (See *Passive Cooling*, page 35).

"Hybrids"

Although solar greenhouses are considered passive solar heating systems, they are quite often combined with components of active—or "moving parts"—systems. One such component might be a rock storage bin connected to the greenhouse and the main house via ducts that use fans and blowers. The rocks inside the bin absorb and store the solar heat until it is needed by the house. The use of "active" elements turns the greenhouse into a *hybrid* system, a combination of active and passive approaches.

Getting It Built

There is a great deal of do-it-yourself potential in the solar greenhouse. Such an approach can save a great deal on labor costs, but the homeowner anxious to get things built should not forget that a successful solar greenhouse requires attention to both design and construction details. The ranks of professional solar designers and builders are growing, so it is possible to have a solar greenhouse project completed by professionals. An intermediate solution would be the purchase of a packaged greenhouse that is assembled on-site. Several of these are on the market (listed below).

The cost of a solar greenhouse depends on design, materials, and labor. A fair estimate of the middle range of cost for a greenhouse is from $3,000 to $7,000. Do-it-yourself projects might cost less and custom design and construction more.

6.
Manufacturers of Greenhouses

J. Scots, Inc
23011 Moulton PKY/B-11
Laguna Hills, CA 92653
(714) 581-9630

Thermodular Designs, Inc
5095 Paris St.
Denver, CO 80239
(303) 371-4111

Sun-Ray Solar Equipment
4 Pines Bridge Rd.
Beacon Falls, CT 06403
(203) 888-0534
Trade name: Sol-Arc

Cottage Industries
RD 2 #129
New Canaan, CT 06840
(914) 533-2125

Weather Energy Systems, Inc
39 Barlows Landing Rd.
Pocasset, MA 02559
(617) 563-9337

Turner Equipment Co., Inc
HWY 117 South
Goldsboro, NC 27530
(919) 734-2351

Solar Resources, Inc
Box 1377
Taos, NM 87571
(505) 758-9344

Advanced Energy Technologies, Inc
Solartown USA
Clifton Park, NY 12065
(518) 371-2140
Trade name: Zerothermic

Vegetable Factory, Inc
100 Court Street
Copiague, NY 11726
Trade name: Vegetable Factory

Lord & Burnham Div
2 Main St
Box 255
Irvington, NY 10533
(914) 591-8800

Abundant Energy, Inc
16 Newport Bridge Rd.
Warwick, NY 10990
(914) 258-4022

Four Seasons Solar Products Corp
672 Sunrise Highway
West Babylon, NY 11704
(516) 422-1300
Trade name: Florida
Trade name: Paradise

Sun Room Co
P.O. Box 265
Reamstown, PA 17567
(215) 267-3864

7.
Solar And Climate

A solar heating system works with climate. The two basic questions are: how much sunshine do you get during the heating season; and how cold is the winter in your area.

For purposes of this discussion, sunlight will be called **solar radiation** (it can also be called *insolation*—meaning *in*cident *sol*ar radi*ation*).

The cold of winter will be expressed in **degree days**.

Microclimate, a third climatic factor, refers to the weather patterns—wind, fog, frost—unique to your site.

Solar Radiation

Solar radiation reaches the earth's surface in such abundance that three days' worth is roughly equivalent to the known reserves of fossil fuels. The **solar constant**, the average solar radiation striking the outer surface of the earth's atmosphere, is 429 Btus per square foot per hour ($Btu/ft^2/hr$). Thirty percent of this is either reflected back into space or absorbed by the atmosphere (see *Figure 1*).

Solar radiation, as it strikes the earth, is either **direct** or **diffuse**. Direct radiation is the bright, sharp sunlight received on a clear day. It casts definite shadows. Diffuse radiation is that which is reflected by atmospheric particles or clouds.

The amount of solar radiation reaching the earth's surface varies greatly on an hourly, daily, and monthly basis (See *Figure*

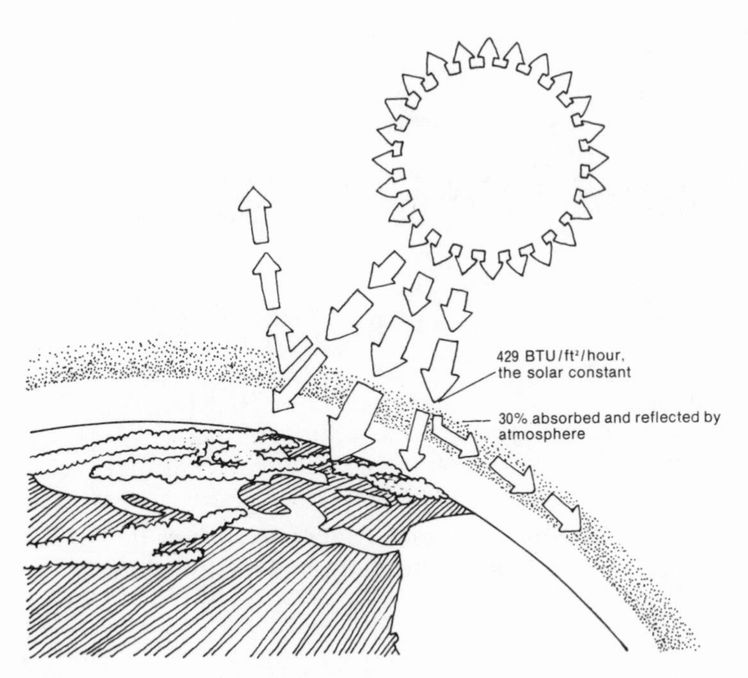

Figure 1. The solar constant: 429 Btus per square foot, per hour

2). The winter sun follows a low path in the sky and its rays strike the earth at a sharp angle, after traveling a greater distance through the atmosphere than the rays do in the summer. The hourly variations in the intensity of solar radiation follow the same principle. In the early morning and late afternoon, the sun is low on the horizon and its rays must pass through more of the atmosphere than they do at noon when the sun is more directly overhead and at its greatest intensity.

The solar radiation reaching any city, town, or village in the United States is determined by latitude and cloud cover. Cities sharing the same latitude can have wide variations in solar radiation because of differing cloud covers.

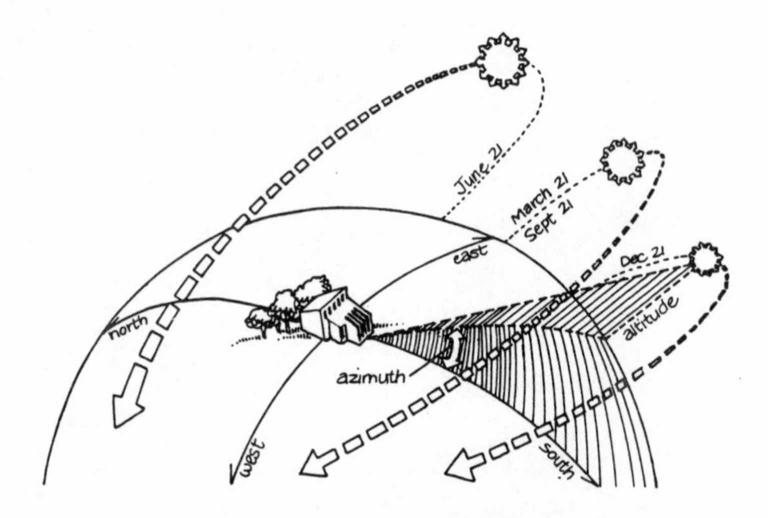

Figure 2. The amount of solar radiation reaching the earth's surface varies on an hourly, daily, and monthly basis

Degree Days

The outdoor temperature of your town is a factor in how well a solar collector located there will perform. The lower the outdoor temperature, the faster a passive solar house loses heat, both through its walls afd through the glass of its collector area. Efficiency and effectiveness are reduced.

Over an entire winter, the daily average temperature is subtracted from a base of 65°F (the temperature below which heat is presumed to be needed), to find the number of **degree days** for that day. So if the average temperature is 40°F, then 40° from 65°F is 25° or 25 degree days for the day in question.

The severity of an entire winter is reflected in the total accumulation of degree days. The more degree days, the more heat a house will require.

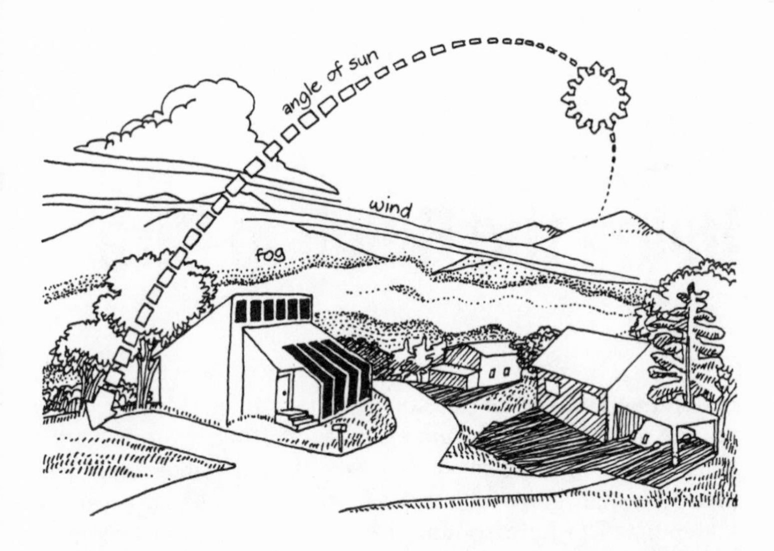

Figure 3. Microclimate is the package of weather factors that are consistent and unique at a particular site

Microclimate

Because solar heating is a "natural" energy source, it is inextricably tied to the natural conditions at the point of its use. Microclimate is the package of consistent weather factors unique to a particular site (see *Figure 3*). Attention to the details of microclimate can greatly improve the performance of a solar project.

For instance, if fog is common in the morning, a solar collection area should not face east of south. Sunlight will be diffuse and weak from that direction. The collection area should be faced slightly west of south to pick up more afternoon sun.

Wind patterns at the site, vegetation, land forms, soil drainage are all points to be considered when assessing a microclimate. By working with the site, rather than against it, you can help a solar house to perform better, and you will therefore be more comfortable.

8.
Using the Building Site

A building site can be developed to help the performance of a passive solar house. The goal is to level off the extremes of the local climate. A number of factors at the site—trees and other plant life, the slope of the land, and soil drainage—can be manipulated to help soften winter's cold and summer's heat.

The ideal location for a solar house is a south-facing slope: access to sunlight is assured and the slope protects the north side of the house from harsh winter winds, which dramatically increase heat loss as they blow against a house. The landscaping around a solar house can help reduce the energy needed to heat and cool it. Evergreen trees or thick (and fast-growing) hedgerows planted to the north are good windbreaks.

In the summer, leaf-bearing trees to the west will shade a house from the setting sun. The sun's position in the sky late in a summer day will make it a very potent source of heat and discomfort if it is allowed to shine directly into west-facing rooms.

The house should be built where the water table is low and where drainage is adequate. Moist soil is a heat sponge, drawing heat away from the building. Dry soil, on the other hand, is a fairly good insulator and can be used for earth berming. (Earth berms are mounds of dirt built up against one or more sides of the house to give added protection against winds and heat loss.)

The most important consideration at the building site for a solar house is access to sunlight.

9.
Access to Sunlight

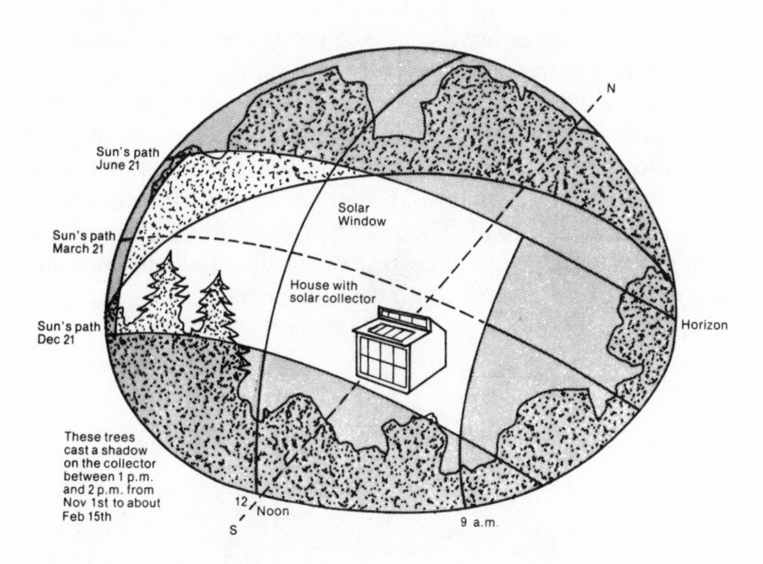

Figure 1. The solar window

The first thing to consider before planning to use solar energy for heating your house, or your hot water, is whether or not you have **access to sunlight**. If you are planning a solar greenhouse or another of the passive heating approaches, then you must have a south-facing wall that is unshaded by trees, land forms, or other buildings during the winter.

The area of access—the "view" from your south-facing wall to the sun—is called the **solar window** (and sometimes *solar skyspace*).

Generally, access to sunlight is considered adequate when 90 percent of the wall or roof area intended for use in solar collection is unshaded during the heating season, between the hours of 9 a.m. and 3 p.m. These are the prime hours for collecting solar energy.

December 21, the shortest day of the year, is when the sun travels its lowest path above the horizon in the sky and consequently casts more and longer shadows. If you get sun on that day, you get it all winter. It's a good idea to monitor the position of the sun for an entire day during the general period of short days in December.

Figure 1 shows how the solar window is formed, by using the sun's paths through the sky on December 21 and June 21, the shortest and the longest days of the year. The east and west boundaries of the "window" are set as the sun's position in the sky at 9 a.m. and 3 p.m. The sun's path on March 21 divides the window roughly in half. *Figure 2* shows the solar window from a point above the north/south-east/west axis.

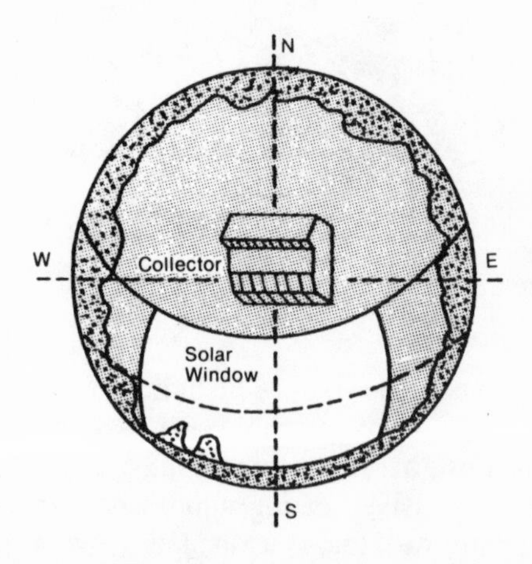

Figure 2. The solar window seen from directly overhead

Figure 1. A south-facing slope is an ideal location for a solar house

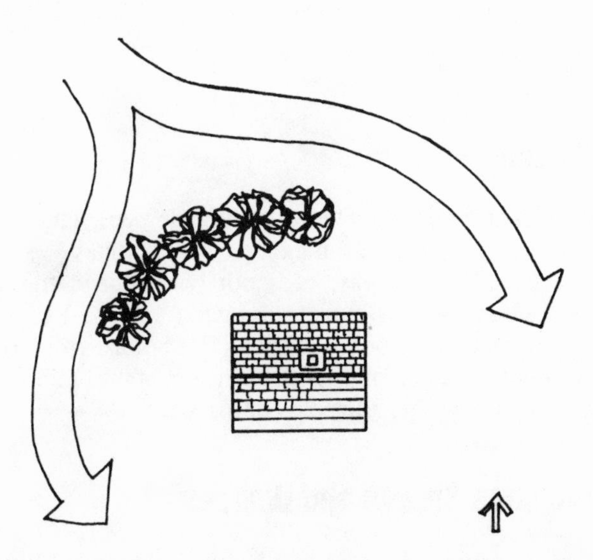

Figure 2. Trees to the north make an excellent windbreak

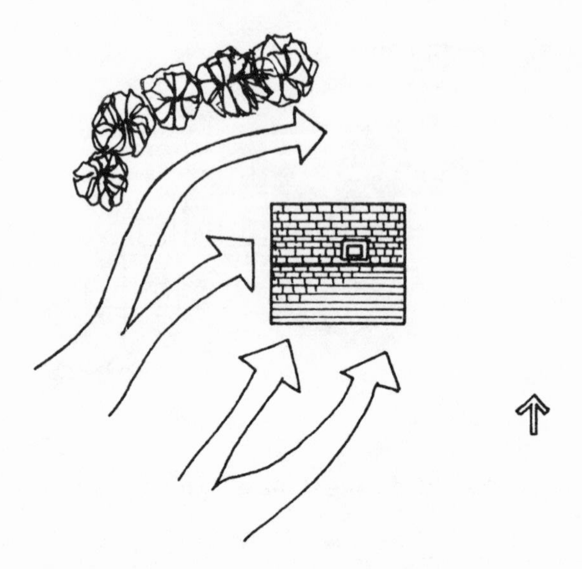

Figure 3. Summer breezes can be channelled to cool a solar house

Orientation

Ideally, a solar house should face within 30 degrees of true south (which is not the same as magnetic or "compass" south). Houses oriented to the east of south will depend more on morning sunlight, just as houses oriented to the west of south will rely more on afternoon sunlight. Variations from south should be made in accordance with the nature of the building, its site, and the climatic peculiarities of the site.

Solar Access Charts and Devices

There are a variety of methods other than the naked eye for checking on access to sunlight. Some are listed below.

Sun Angle Charts

Bennett Sun Angle Charts
6 Snowden Rd.
Bala Cynwyd, PA 19004

Koolshade Corp. Sun Angle Charts
1705 Gardena Ave
Glendale, CA 91204

Lof Corp Solar Angle Charts
811 Madison Ave
Toledo, OH 43695

Mancock Comprehensive Design Sun Angle Charts
Box 4192
Stanford, CA 94305

Others

Shadow Study Kit
IN-AR Faculte de L'Amenagement
Universite de Montreal
5620 Darlington
Montreal, Que., Canada
Set of 12 diagrams and maps to be used with a light bulb; determines how much sunlight any part of a building will receive.

Sun Angle Calculator
Libbey-Owens-Ford
Toledo, OH 43695
Set of curves and plastic overlays used to determine solar angles; intended for window design.

Sun Angle Desk Top Calculator
Zomeworks Corp.
Box 712
Albuquerque, NM 87103
Calculations that provide hours of sunlight for the days of the

year, elevation in degrees from the horizon, and compass bearings from south.

Sunbloc
Pacific Sun, Inc
540 Santa Cruz Ave.
Menlo Park, CA 94025
Precision site assessment package, including a Brunton type hand transit, plotting sheets, and an instruction book.

Sun Scope II
Solar Usage Now, Inc
PO Box 306
Bascom, OH 44809
Instrument that permits sun sitings for any hour in each season.

Solar Site Kit
Solar P.I.E.
P.O. Box 506
Columbus, NC 28722
Tilt calculator, sun site setter, compass, and notebook

Solar Site Selector
Lewis Associates
105 Rockwood Drive
Grass Valley, CA 95945
Sophisticated instrument that calculates total hours of sunshine, sun paths, and hour segments.

10. Passive Cooling

While a passive solar house warms itself in the winter by collecting and storing solar heat, it also needs to stay cool in the summer. Passive cooling can be as valuable an asset as its winter counterpart because it can reduce or even eliminate the need for expensive air-conditioning.

The structure, layout, and materials of the house are the components of a passive cooling "system," just as they are for passive solar heating. Passive cooling will be most effective in new houses intentionally designed to take advantage of it, but many of the basic principles can be applied to existing houses to make them more comfortable.

Keeping Heat Out

The planet Earth exists in a state of thermal equilibrium, losing as much heat to space as it gains from the sun. Each morning the sun rises and begins to heat up the atmosphere and surfaces of the earth. One of the fundamental purposes of any house is to put a little distance between itself and the earth's daily routine of gaining and losing heat. The first step toward keeping a house cool while the temperature outside is rising is to prevent direct summer sunlight from entering through windows.

Passive solar houses make use of roof overhangs to shade their large south-facing windows. These overhangs are sized to block the rays of the high summer sun, but they should, in the winter, allow the sun to enter and warm the house. An overhang should also be vented so that heat does not build up underneath it. Movable awnings are another simple approach to shading windows. Movable insulation, either interior or exterior, is commonly used to prevent heat loss on winter nights. (See *Movable Insulation*, page 40). These insulating shades, shutters, or panels can also be closed to prevent heat gain during summer days. The appropriate use of shade trees is another part of passive cooling strategy. It is largely a misconception that deciduous trees will graciously admit winter sun through their bare branches in winter and then, while fully leafed, provide shade from the summer sun. To provide actual summer shade for a south wall the trees must be virtually on top of the house, which means that the branches and some of the trunks will block some winter sunlight from the house. However, trees are very good for shading the east and west sides of the house from the summer sun in its low positions as it rises and sets. The low late afternoon sun is especially potent for overheating west-facing rooms. Summer shade trees are important in these strategic positions.

Another method of reducing undesirable summer heat gain is the use of light-colored or white roofing material to reflect sunlight. This effect is also useful on the east and west walls of the house. If heat gain can be controlled, it can simplify other passive cooling tactics, making them more effective.

Letting the Breezes Through

A house can be designed to channel summer breezes to maximize the effect of natural cross-ventilation. The movement of air has a soothing effect on overheated human beings: a gentle breeze must be nature's apology for its harsher summer behavior.

The best way to take advantage of natural ventilation is to provide a clear path for air movement through the house. Many conventional houses, like passive solar houses, allow for this movement, only requiring that you activate it at the right time of the day. For instance, if the house is allowed to take advantage of cool night brezes, the interior should be quite cool by morning. Throughout the morning and possibly into the early afternoon, the house will probably remain cooler than the outside, if heat gain is prevented.

This will last longest in those passive houses that have thick storage walls and floors (thermal mass) and that are well insulated. The storage mass, cooled by ventilation throughout the night, will absorb heat from the indoor air by day; the insulation will prevent heat transmission through the walls and roof to the living space.

During the hours that the house remains cool, ventilation might actually warm the house by bringing in warm outside air. Sometime during the day, however, the interior space will begin to warm up as warm outside air begins to infiltrate, and as heat is conducted inside (heat is relentless in its search for equilibrium). That is when you should open up the path of cross-ventilation. In a well-designed passive house, the designer will have paid careful attention to the layout of rooms and to the characteristic weather patterns of the building site. In attempting to link the house to its site, the designer can use earth berms, vegetation, or wing walls to channel breezes. You can then open the necessary windows, doors, or vents when you feel it is appropriate. Just as people did in the days before the air-conditioner, you will learn to manage and take advantage of the assets of your home's design.

Inducing Ventilation

Another approach to passive ventilation makes use of the principle that hot air rises. Any good design will automatically include roof vents to allow heat that builds up in attics or upper floors to escape. By creating a "solar chimney," an effect where heat is allowed to rise rapidly and vent itself, air will be drawn up through the house. This induced chimney effect, however, calls for the controlled entrance of replacement air into the lower part of the house. In other words, as the hot air is rapidly vented, cool replacement air should be drawn in. This requires a source of cool air, which can be something as simple as a well-shaded north yard. A more sophisticated method is to draw the air through buried "earth" tubes; these cool the air by allowing it to give up heat to the ground, which maintains a steady temperature range between 45° and 55°F.

Cooling By Night Sky Radiation

Night sky radiation is a challenging and fascinating phenomenon. The classic passive solar use of the concept is found in Harold Hay's patented Skytherm system. In Hay's system a water pond is installed in the roof of the house. Its purpose is to provide both heating and cooling. During the winter the water is exposed to the sun to be warmed so that it will radiate heat into the living space. At night an insulated cover is rolled into place to prevent heat loss. (The cover slides over the garage during winter days).

In the summer the routine is reversed. The water bags are exposed at night so that they can radiate their heat to the sky—a process that is most effective when there is no cloud cover and the day-to-night temperature difference is at least 20°F. The well-chilled bags are then covered during the day as they draw heat from the living space. The cover is removed again at night and the cycle is repeated.

Night sky radiation is most effective on clear nights and, by extension, it will make more sense in areas that experience a good many clear summer nights.

Night sky radiation is somewhat impractical in many building situations, but innovative designers are challenged to find ways of adapting its use to standard building techniques.

In Summary

A well-designed passive solar house is one that both heats and cools itself, passive cooling being in some cases no more than a clever orchestration of commonplace building elements. As long as there are shades, shutters or awnings, white roofing materials, shade trees, summer breezes, appropriately placed and managed windows, and cross-ventilation, a house has a good chance of remaining comfortable without the expense of air-conditioning. The ability to determine accurately the right combination of cooling techniques for any climate and building site will continually advance as the state of the passive cooling art progresses. Passive solar design will be more widely recognized as a means of effectively providing year-round comfort and controlling home energy costs.

11. Movable Insulation

The large glass areas of a passive solar heating system let a great deal of solar energy in, but they also let a lot of heat out. On a sunny day much more heat is gained than is lost, but at night almost all of the heat flows out. The various kinds of movable insulation—insulated shutters, panels, curtains, and shades—are designed to minimize this heat loss by covering the glass at night.

In some northern parts of the country, the use of movable insulation can double the heating contribution made by a passive solar system. Or it can extend the usefulness of a smaller system. Either way, the money savings are substantial. Conventional, well-insulated houses typically lose 20 to 30 percent of their heat through windows. Movable insulation can reduce these losses by as much as 90 percent.

Homeowners using movable insulation report that their houses are more comfortable. They don't feel drafts; the room air feels warmer; the thermostat can be set lower without discomfort.

Some movable insulation uses reflective surfaces on the window side so that it can be used on summer days to block out the sun. This lowers electricity bills for air conditioning.

It is not worth the cost to invest in movable insulation or solar heating until the major areas of heat loss in a house have been taken care of. The first step is to caulk and weatherstrip any leaks around window, doors, electrical outlets, etc. (Movable insulation should not be used to stop drafts from leaky windows and frames.) Insulating the attic (and walls, if possible) should be done next.

Cost

Most commercially available movable insulation costs $3 to $8 per square foot, or $60 to $150 for a 3 by 6 foot window. Some of the larger, well-finished wooden shutters and sliding panels can cost $350 per window or more. Cost can be minimized if such devices are planned as part of the initial construction or renovation of a house or room. Panels can be made to slide into adjacent walls when not in use; window frames can include tracking channels for shades; access to curtains and shades can be assured in front of passive thermal storage walls.

Though outdoor shutters and shades can be just as effective, it is usually cheaper and more convenient to use indoor movable insulation. It can be applied and removed without going outside. Indoor shutters also can be much lighter in weight than outdoor shutters, which need to be fortified against the weather.

How It Works

The typical window absorbs heat from the house and transfers it to the outside at a fairly constant rate (depending upon the outside temperature and wind conditions). Movable insulation keeps the glass from "seeing" the heat radiating from the warm objects and walls of a room, and does not allow warm air currents to pass in front of the window. Instead, it places "still air" in front of the window. The best insulators, made of foam, contain thousands of tiny air pockets separated by very thin material of low conductivity.

Making It Effective

The do-it-yourself homeowner armed with an understanding of how movable insulation works can attack the task of fabricating movable insulation, and spend time instead of money. Whether the device is homemade or a commercially available product, three problems must be solved to make movable insulation effective: sealing the cracks between the insulation and the wall or window, providing a vapor barrier, and attaching the insulation to the window or wall.

Condensation may occur when warm inside air comes into contact with cold glass. This increases heat loss and may stain or rot sashes and frames. By keeping heat in, movable insulation makes the inner surface of the glass even colder, increasing the risk of condensation when warm air strikes it. A good seal reduces condensation *and* heat loss. Sealing the insulation can be done in two ways. An **edge seal** is one in which the insulation is sealed only at the edges. This provides another air space between the insulation and the glass and allows the glass, sash, and often the jamb to benefit from the insulation. Almost all insulated shutters, curtains, and shades are equipped with seals to close the cracks between the edges of the insulation and the window or wall. Materials like compressible foam weatherstripping work well as sealing agents for the edges of shutters and panels. Curtains often have magnets or weights in the bottom to make a seal against the window sill or floor. A valance can seal the top. Magnetic tape, Velcro®, staples, hinged wooden strips, compressible foam, and tracking channels are used as side edge sealers. Edge-seal shutters, panels, and shades must be measured precisely; window frames (especially older ones) often are not perfectly square. A panel that does not quite fit loses most of its effectiveness and causes condensation problems.

This is not true for a **face-seal** or glass-hugging panel, in which the insulation is pressed directly against the glass with the sash left uncovered. A face-seal panel that is cut slightly too small will be almost as effective as a perfectly fitting one. The small amount of air between insulation and glass, if they are less than ½ inch apart, moves very slowly, producing little or no condensation on the glass.

Condensation may also occur within the insulation. A **vapor barrier** will prevent this. A sheet of absolutely hole-less polyethylene plastic, foil or vinyl makes a good vapor barrier for curtains, shades, and some kinds of shutters and panels. Extruded foam insulation, the kind called "closed cell," is its own vapor barrier. *Attaching* the insulation is a consideration for pop-in panels. If magnets, Velcro, or similar materials are used, a silicone sealant or urethane bond adhesive can hold these attaching materials to the glass. A friction fit into the window frame or recess using compressible foam or a heavy fabric can also work.

Comparing Products

Along with cost and appearance, the following factors are bases for comparing the various types and brands of movable insulation.

R-Value. A material's resistance to heat flow is expressed as its R-value. The higher the R-value, the more it slows the flow of heat. However, there is a point beyond which the cost of a higher R-Value may outweigh the benefits unless, for example, the higher-R product is the only one that will work in a given situation. Generally, insulation with an R-value no lower than 4 or 5 (66 to 71 percent heat loss reduction) is recommended for movable insulation being used as part of a passive heating system. The warm thermal mass of such a system is usually close to the glass and radiates unusually high amounts of heat.

Durability and Safety. Most wear occurs at the edge seal of a movable insulation device. Shades fray at the edges; tracking channels break, tear, or buckle; compressible foam tears or wears away. Replacement should be planned for.

Some of the lightweight, high R-Value types of foam insulation are flammable, spread fire very quickly, and may release toxic gases when they burn. Know what the properties of your insulation are. If such insulation is used, make sure it is used in accordance with fire safety codes.

Convenience. A shutter, panel, curtain, or shade that is not used prevents no heat loss, no matter how high the R-value is. Movable insulation that requires little attention and is lightweight will tend to be used more than other, less convenient kinds. Some products cover and uncover the glass on a signal from a timer or in reaction to the amount of sunshine, but these tend to be relatively expensive. Most insulated shutters, curtains, and shades are convenient in that they remain at the window site whether opened or closed. Removable panels are the cheapest and easiest to make, but require storage space.

Movable insulation can be simple or complex, subtle or the center of attention. It is essential in a passive solar heating system and usually economical in conventional houses. Homeowners can save money and energy, simply by closing the curtain on heat loss.

12. Ancient Techniques of Passive Solar Design

The elements of what we call "passive solar design" are found in the ruins and writings of ancient Greece, Rome, China, Egypt, and the Americas.

Many solar designs now being introduced are re-inventions of techniques put aside in times of plenty or lost in times of upheaval.

Orientation

This, the first principle of solar design, dictates that a house be built to face the winter sun. In the northern hemisphere, this means placing most (if not all) windows on the south side. The sun's path in winter is a low arc through the southern sky: a south-facing house will collect available sunlight while losing less heat through its north, east, and west sides. In summer the sun rides a high arc from east to west which carries it briefly into the high southern sky. A south-facing house with a roof overhang shields its inhabitants from direct summer sunlight.

Glass Windows

A glass window does more than merely admit sunlight into a room: glass traps the infrared heat reradiated by objects in the room. This trapping is known as the "greenhouse effect." Windows were not always made of glass. The ancient Chinese used silk, rice paper, and oyster-shell linings. The ancient Romans used mica and selenite. We know from excavations at Herculaneum and Pompeii, and writings of Pliny and Seneca, that after the Romans found the secret of manufacturing glass in large panes, they used it for the solar heating of the public baths and the sun rooms and greenhouses of the wealthy.

Thermal Mass

The capacity to absorb the sun's heat by day and release it gradually in the cool of night distinguishes a functional passive solar home from a house with south-facing glass. It was for its great heat-storage capacity (and availability) that adobe (sun dried brick) was a favorite building material of many early peoples: Greeks, Romans, Egyptians, Persians, American Indians. In contemporary passive solar structures, thermal mass may take the form of masonry, earth, or water.

Ancient Greece

As the ancient Greeks increased in number, so grew their demand for wood. With wood they cooked their food, heated their homes, built their ships. They stripped their forests for wood, leaving the ground bare of trees and defenseless against erosion. While city-state governments responded to the shortage of wood by taxing its sale and restricting its use, architects adopted climate-conscious design of houses and even entire new towns.

The city of Olynthus, for instance, built an addition in the fifth century B.C. according to Hippodamus's checkerboard plan— a rectangular grid that allowed each building to face south and thereby receive the most possible winter sunlight. The east-west avenues were built wide so that all houses were open to the southern sky. The south side of each house had an open portico with earthen floors and adobe walls to collect sunheat, an overhang to provide shade from the sun at its midday summer height, and a low wall to keep out cold ground-level night drafts.

The builders of the city of Priene adapted the checkerboard to the shape of the city's new site—the steep south side of Mt. Mycal. They ran north-south stairways up the mountainside and terraced east-west avenues along the mountain contours.

Socrates, explaining the connection between usefulness and beauty, advocated solar design thus:

> When houses face south, in the winter the sun lights the inner room and in the summer makes shade because its path is directly over us and the roof. If, then, it is good to have the house this way, we must build the southern side higher, so that the winter sun is not blocked out. The north side must be lower, so that the cold winds don't blow in. To put it briefly, the house in which a man finds the most pleasant shelter at all seasons and which can keep his possessions safest is the house that is presumably the most pleasant and the most beautiful.

Ancient Rome

The demand for wood only increased with the passage from Greek civilization to Roman Empire. By the 4th century after Christ, the Romans were so lacking in fuel that the government commissioned a fleet of ships to import wood from faraway France and North Africa. Architects such as Vitruvius, Faventinus, and Palladius began to devise energy-saving house designs. By their counsel, wealthy Romans had their villas built with southern orientations and with thermal mass under their sun-lit dining room floors: a dark mix of rubble, broken earthenware, sand, ash, and lime. In place of the wood-consuming *hypocaust* central heating, they heated their villas with glassed-in sun rooms, *heliocamini*. Ulpian outlawed the building of structures that put existing heliocamini in shadow. This ruling, the first solar access law, was incorporated into the Justinian Code in the 6th century.

Pueblo Indians

Against the fierce day-to-night temperature swings of the American Southwest, the Pueblo Indians of the 11th and 12th centuries built their dwellings with heavy walls of adobe, mud, and stone. To minimize the surface area exposed to the outside, the Pueblos put many dwelling units into large community structures, ruins of some of which can be seen today in New Mexico and Arizona: Acoma, Mesa Verde, Montezuma's Castle, Pueblo Bonito.

Ancient China

Traditional Chinese architecture shows a manifest climate conscious design comparable to that of the ancient Greeks. Streets of important cities are laid out on rectangular grids, aligned with the compass; houses and temples face south, the direction of summer, warmth, and health. The typical house is built around a south-facing courtyard. All openings face inwards. Wide decorative hangings provide summer midday shade and throw off heavy monsoon-season rainwater.

Europe

When Rome fell, so fell demand for and supply of glass windows. In the anarchy of the Dark Ages, homes had to serve as small fortresses. Windows were too vulnerable. Greenhouses disappeared also—the church forbade them, deeming it sacrilegious to grow plants apart from natural habitat and season. It was not until the later renaissance of the 16th century that greenhouse horticulture revived. With the discovery of the New

World came exotic foreign plants and new wealth; with the discovery of new worlds through Galileo's telescope came empirical challenge of Church doctrines. The renewed spirit of innovation was felt in numerous solar design developments, up through the present century.

- The Italians of the early Renaissance revived the Classical styles of architecture and, with time, Vitruvius' solar orientation principles. The northern Europeans later copied the forms but neglected the solar orientation that made them functional.
- A thermal-mass technique known as the "fruit wall" came into fashion among French and English horticulturalists of the 17th century. They found they could extend a fruit vine's growth into the cooler seasons by fastening it to a sun-lit wall, usually of brick.
- The 18th century saw the publishing of treatises on the theory and construction of greenhouses: Pieter Van de Voort's *Landhuren, Lusthaven, Plantagian* (Dutch, 1737); Michael Adamson's *Famile des Plantes* (French, 1763).
- Conservatories—lavish greenhouses designed to give pleasure rather than heat or food—came into fashion among wealthy 19th century Europeans, especially the British. However, conservatories came to be built for the use of artificial heat rather than solar heating as fossil fuels became cheaply available. When these fuels were rationed during the First World War, those British conservatories that lacked solar design became unsupportable luxuries.
- Solar access was denied to the vertically crowded tenement dwellers of the Industrial Revolution. Now and then, a developer would engage in reform. In the 1860s, the Lever Brothers Company built Port Sunshine for its Liverpool soap factory workers. Although the plan of this community did not give the houses a consistent southern orientation, it did leave plenty of open space for solar access.
- The solar housing development of Neubhl was built near Zurich after the First World War. The houses were well oriented, with south-facing bedrooms and north-facing kitchens. They were spaced well apart to insure solar access. The shades and awnings over their windows were retractable.

- In Germany during the Weimar period, progressive architects such as Walter Gropius built housing developments with what they thought would be good solar orientation—windows open to the east to admit the morning sun and to the west to admit the afternoon sun. The resulting heat gain in these *Zeilenbau* ("row house") apartments was small in winter and great in summer— uncomfortable either way. An architectural movement followed which did face apartments southwards. But as those housing developments were intended to be worker communities, the Nazis branded the movement "Communistic" in the 1930s and it fell apart.

America

Solar design began to receive attention not only from the press but from industry. The Libbey-Owens-Ford Glass Company marketed the first double-paned glass for solar houses in the early 1950s. Two layers of glass sandwiching a hermetically sealed space of dead air had twice the thermal resistance of a single pane.

The designer's question of how much glass to put in a solar home was addressed by F.W. Hutchinson, a professor of mechanical engineering at Purdue. Hutchinson found a key to controlling over- and under-heating of a sun-lit room in the ratio of glass area to floor area. He produced tables from which solar architects could calculate the best ratio for a given climate.

The new interest in solar design shown by American home buyers in the early 1940s waned by decade's end. Many homes billed as "solar" turned out to be ineffective—due to improper orientation, lack of proper insulation and thermal mass, and the use of single-paned rather than double-paned glass. As the price of electricity and fossil fuels fell after the second World War, even the best-designed solar home's performance could no longer compensate for its higher initial cost. Electric and fossil-fuel heaters—automatic, mass-produced, inexpensive—were becoming popular. The wartime conservation ethic was yielding to a consumption ethic.

Nonetheless, research continued at several universities. The Massachusetts Institute of Technology built four houses over the period 1939 to 1956 that were heated by solar design, both passive (south-facing glass with overhangs for summer shade) and active (collectors, fans, air- or water-circulation). One of M.I.T.'s solar experimenters, Dr. Maria Telkes, explored a new type of thermal mass: a hydrated salt that melts at temperatures close to room temperatures when heated and recrystalizes releasing stored heat when cooled. In 1949 Telkes used such a *phase-change* material, popularly known as Glaubers salt, to store heat in an experimental house in Dover, Masachusetts.

Another thermal mass technique was introduced in 1956 by two Frenchmen, architect Jacques Michel and Dr. Felix Trombe (see page 4), director of solar energy research for the Centre Nationale de la Réchêrche Scientifique at Odeillo, in the Pyrenees. In their scheme, a concrete wall placed several inches behind a south-facing glass wall created a space in which air is heated quickly by the sun. So heated, the air circulates vigorously through vents cut in the top and bottom of the concrete. A rock bed can be placed underneath the wall to increase the amount of thermal mass. This arrangement, now popular as the *Trombe Wall* (see page 4), was anticipated by the "Mass Wall," an 1881 patented design by Professor E.L. Morse of Salem, Massachusetts.

Not all progress in solar design has come from large institutions. Solar design—especially the passive variety—has been a practical focus for independent-spirited individuals. One spawning ground for do-it-yourself solar creativity has evolved in Arizona and New Mexico over the past quarter-century, seeded by innovators such as Steve Baer, Peter Van Dresser, and David Wright. In the land where the Pueblos built their sun-dwellings a thousand years before, the self-reliant solar home-builders of the present day pride themselves on common sense, climate consciousness, and independence from government bureaucracy and corporate control—even as solar researchers at the nearby Los Alamos Scientific Laboratories study theoretical energy flows through imaginary buildings simulated by computer.

13. A Career In Solar Energy

The solar energy industry, though still young, is quite diverse, and it offers a number of avenues by which to pursue a job or career. The clearest potential is for professionals and tradesmen who can work in the solar heating field—designers, builders, or installers of solar heating or hot water systems.

There are four basic groups within the professions and trades of the solar heating industry:

- Solar architects and designers use their architectural and solar design background and their building experience to design solar buildings and additions;
- Solar designer/builders use their design and building experience to construct solar buildings and additions.
- Solar installers combine skills of roofing, plumbing, sheet metal work, electrical work, and carpentry to install active solar hot water systems in buildings;
- Solar consultants advise on aspects of passive or active solar systems, including design, financing, economic feasibility, and sizing.

Someone already at work in the design profession or building trades will be best prepared to make a transition to a solar oriented career. Work in the solar heating field is not something new and separate; it is usually a new *specialization* within an existing field. This does not mean that someone starting fresh in an established field could not do so with the full intention of specializing in solar. But a solar builder must first be a builder.

Solar and solar-related opportunities also can be found in research and development, manufacturing, and distribution. There are opportunities for scientists, engineers, managers, technicians, many types of skilled and unskilled laborers, salespeople, marketing experts, etc. Such fields as finance, investment, real estate, insurance, and law also need solar specialists.

Solar Training

Whether it be academic or vocational, some form of solar training is necessary both for novices and for established professionals and tradespeople. An increasing number of colleges, universities, technical schools, community groups, solar manufacturers, utilities, and trade unions offer solar courses or training programs.

Consult the *Sources of Information* section on page 137 for an agency or organization in your state that can provide a list of solar educational or training programs in your area.

Typical Required Training

Architect: Building design; passive system design.

To work: Bachelor, architecture; master, architecture; or equivalent informal training.

To work independently: three or more years experience and state license exam with Bachelor degree; two or more years and state licence with masters; twelve years plus licence with informal training.

Mechanical Engineer: Thermal design; specs for construction details of active and passive systems (including HVAC); performance analysis.

To work: Engineering degree or informal training.
To work independently: With degree, 4 years of experience plus state E.I.T. and P.E. exams; without it, 12 years experience plus state E.I.T. and P.E. exams.

Solar System Designer: Sizing; components choice; specs; working drawings and details for all solar systems.
To work: Informal training.

Building Designer: Design of private buildings below a specified price.
To work: Informal training.

Structural Engineer: Building design
To work: Engineering degree or informal training.
To work independently: With degree, 4 years experience plus state E.I.T. and P.E. exam; without it, 12 years experience plus exam.

Environmental Designer: Landscaping, passive building and site design.
To work: Informal training.

Electrical Engineer: Electrical system and controls design; research and development.
To work: Engineering degree or informal training.
To work independently: With degree, 4 years experience plus state E.I.T. and P.E. exam; without it, 12 years experience plus exam.

Chemical Engineer: System design; fluid characteristics and flow calculations; heat transfer calculations; corrosion studies; performance analysis.
To work: Same as for electrical engineer.
To work independently: Same as for electrical engineer.

Interior Designer: Passive building design.
To work: Informal training.

Installer: Installation and maintenance of systems.
To work: Contracting experience or formal training.
To work independently: HVAC or master plumber status; licence in a few states.

Builder: Oversee construction of active and passive building projects.
To work: Informal training.

HVAC contractor: Installation and maintenance of active systems.
To work: Informal training.
To work independently: Licence or master plumber status.

Carpentry Contractor: System installation and construction.
To work: Informal training.

Masonry Contractor: Passive building.
To work: Informal training or apprenticeship.

Glazier: Installation of glazing for passive systems.
To work: Apprenticeship.

Mechanical Contractor: Installation and maintenance of active systems.
To work: Informal training.
To work independently: Master plumber status.

Roofing Contractor: Installation and maintenance of active air systems.
To work: Apprenticeship.

Electrical Contractor: Installation and maintenance of electrical wiring and controls of systems.
To work: Apprenticeship.
To work independently: Local licence.

14. Solar Construction Details

Although passive solar heating is simple in concept, the details of a passive solar house must be executed properly if the project is to be successful. The construction drawings that follow are intended to inform a prospective passive solar home owner about some of the nuances involved in the building of such houses. They can also be used when sitting down with a builder to discuss a project, to help the client gain confidence in what he or she wants.

The detailed drawing of the solar greenhouse on page 16 can be used in the same way. Note that the harmonious relationship between the five passive elements (collector, absorber, storage, distribution, and control) is continually emphasized.

Direct Gain

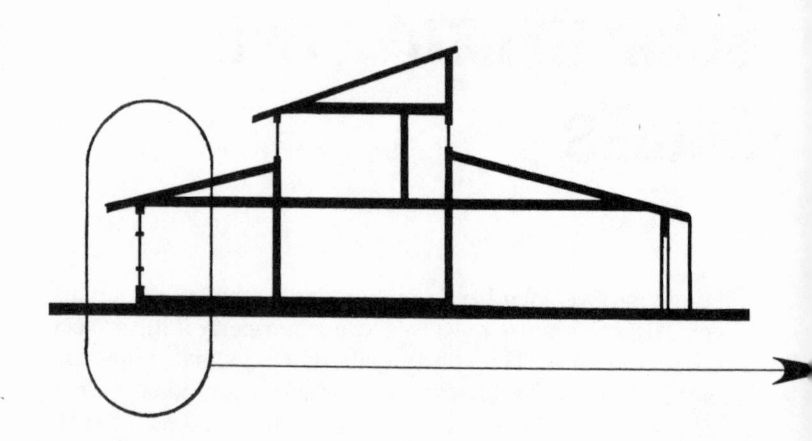

Direct gain construction details

1. Collector: glazing in one to three layers, as appropriate to climate.
2. Absorber: dark colored surface (tile, brick, or concrete) exposed to direct sunlight with minimal coverage by carpet or furniture.
3. Storage: concrete slab floor in direct contact with absorber surface; insulation below.
4. Distribution: re-radiation of heat stored in concrete floor; convection.
5. Control: overhang sized to provide full shading of collector glass at summer solstice (June 21).
6. Control: automatic or manual roll-down insulation.
7. Control: high and low vents for summer ventilation, protection against overheating.

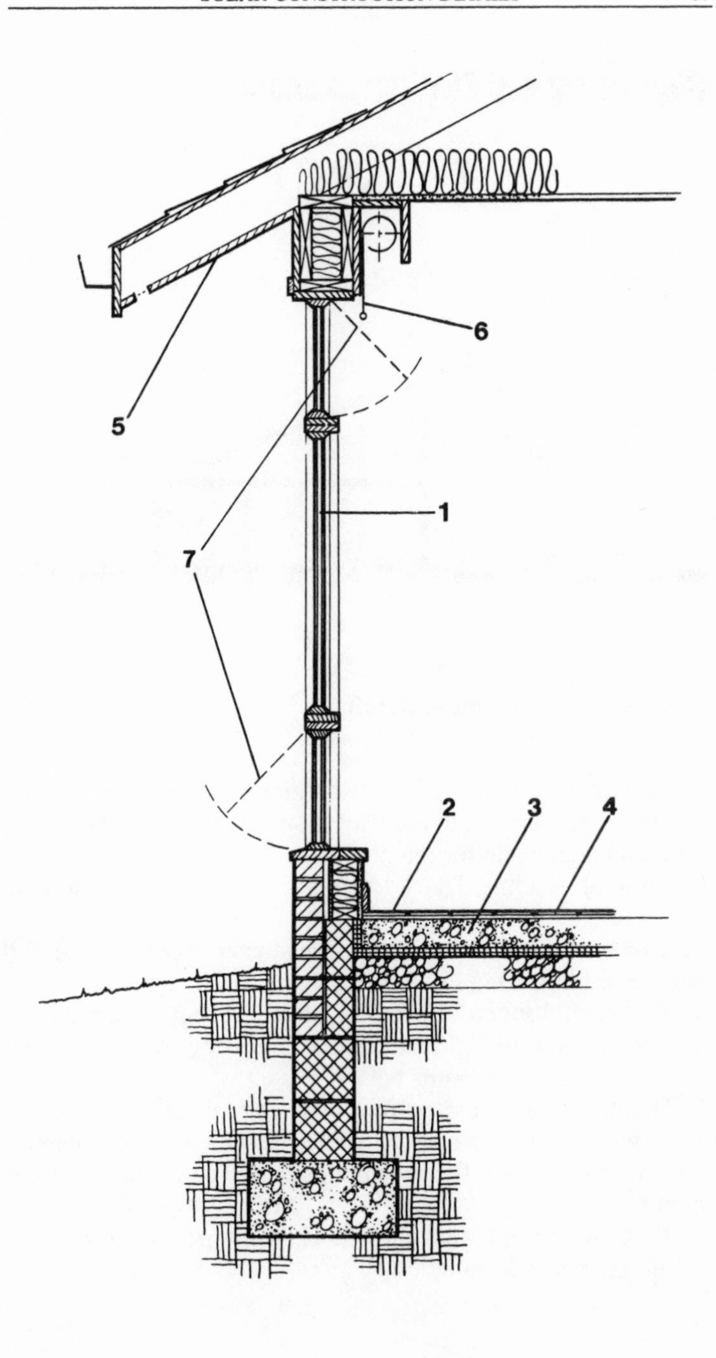
6
5
1
7
2
3
4

Clerestory and Skylight Insulation

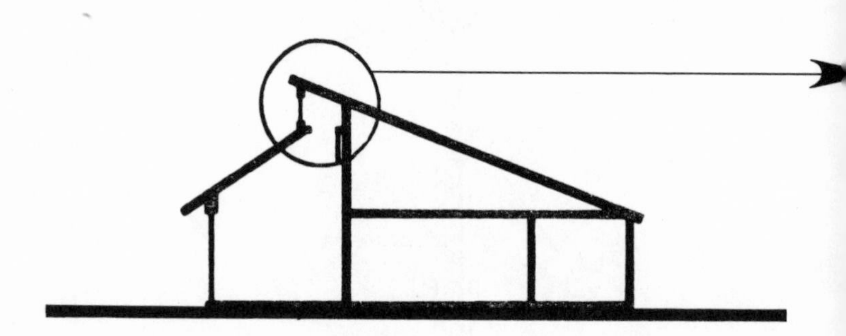

Clerestory construction details

1. Collector: (glass, acrylic, fiberglass, etc.) in one or two layers (as appropriate to climate) mounted to be air- and water-tight and accessible for maintenance.
2. Control: shading device to block all sun at summer solstice (June 21).
3. Conservation: compression weatherstripping around full perimeter minimizes losses of stratified warm air.
4. Control: hinged insulating panel swings down during daytime, admitting light and warmth, and upward during night to conserve heat in rooms below.
5. Distribution: opening from clerestory ridge to outside allows hot stratified air to be vented during summer months; opening to be sealed tightly with insulating panel during winter months.
6. Control: pulley and cord mechanism permits control of insulating panel from below.

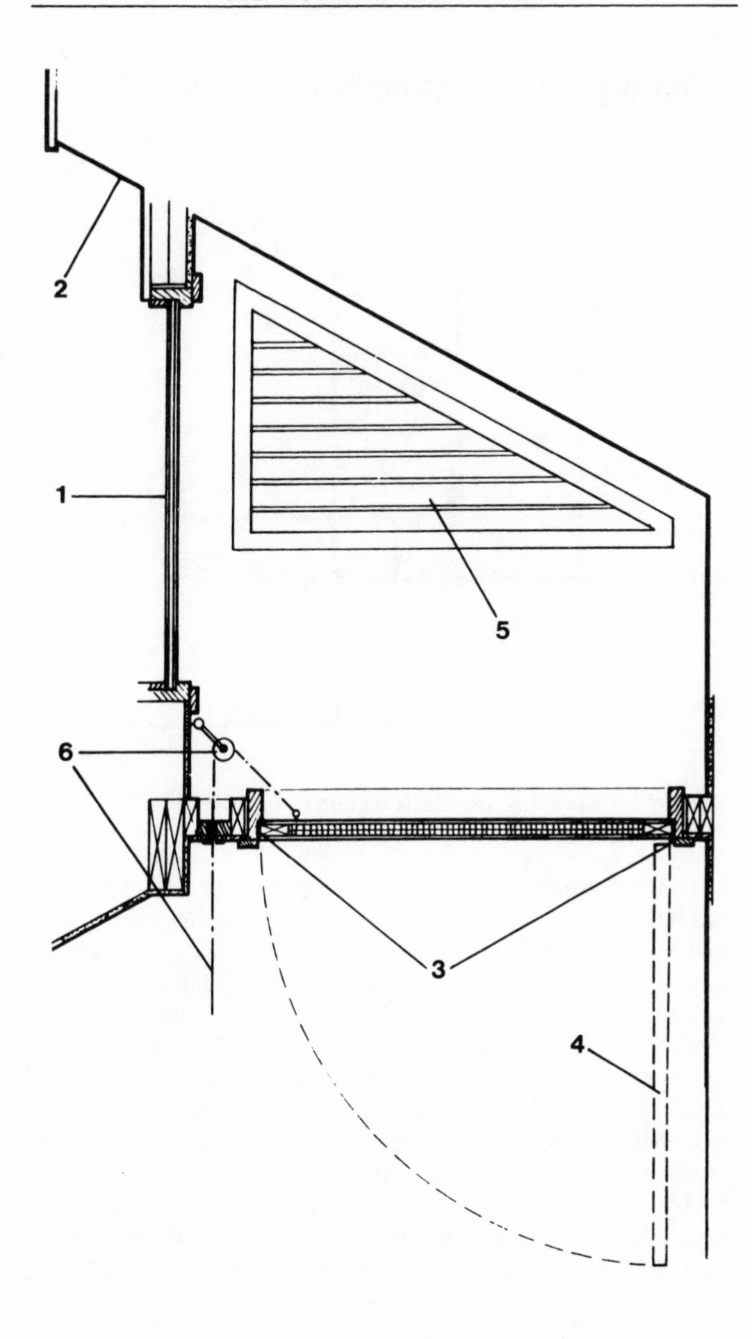
2
1
5
6
3
4

Bi-Folding Movable Insulation

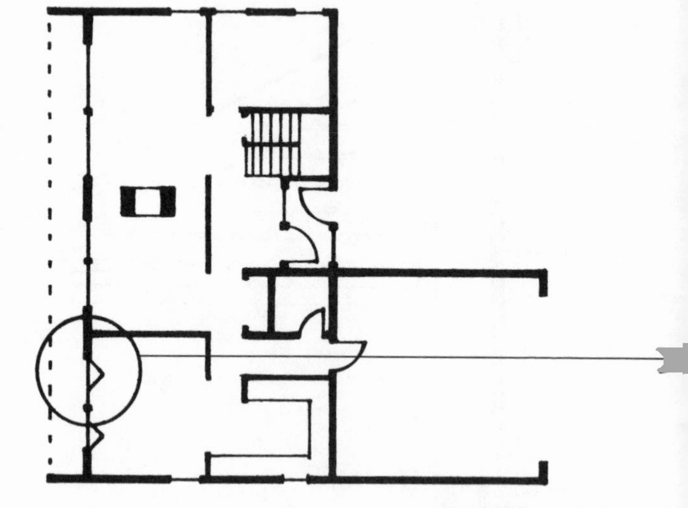

Bi-folding movable insulation, construction details

1. Conservation: compression weatherstripping at jamb, sill, header, and where two sections of bi-fold panel meet, minimizes infiltration.
2. Construction: piano hinges securely fastened allow the hanging of a trackless panel and result in panel storage flush to adjacent wall; track-mounted systems open 90° only.
3. Conservation: latches at floor and head; panels lock and seal to minimize perimeter infiltration.
4. Control: movable insulating panels over glass reduce winter-night heat loss and summer-day heat gain.
5. Control: space between glazing and movable insulation should be vented to exterior by a top opening, to avoid potential summer overheating.

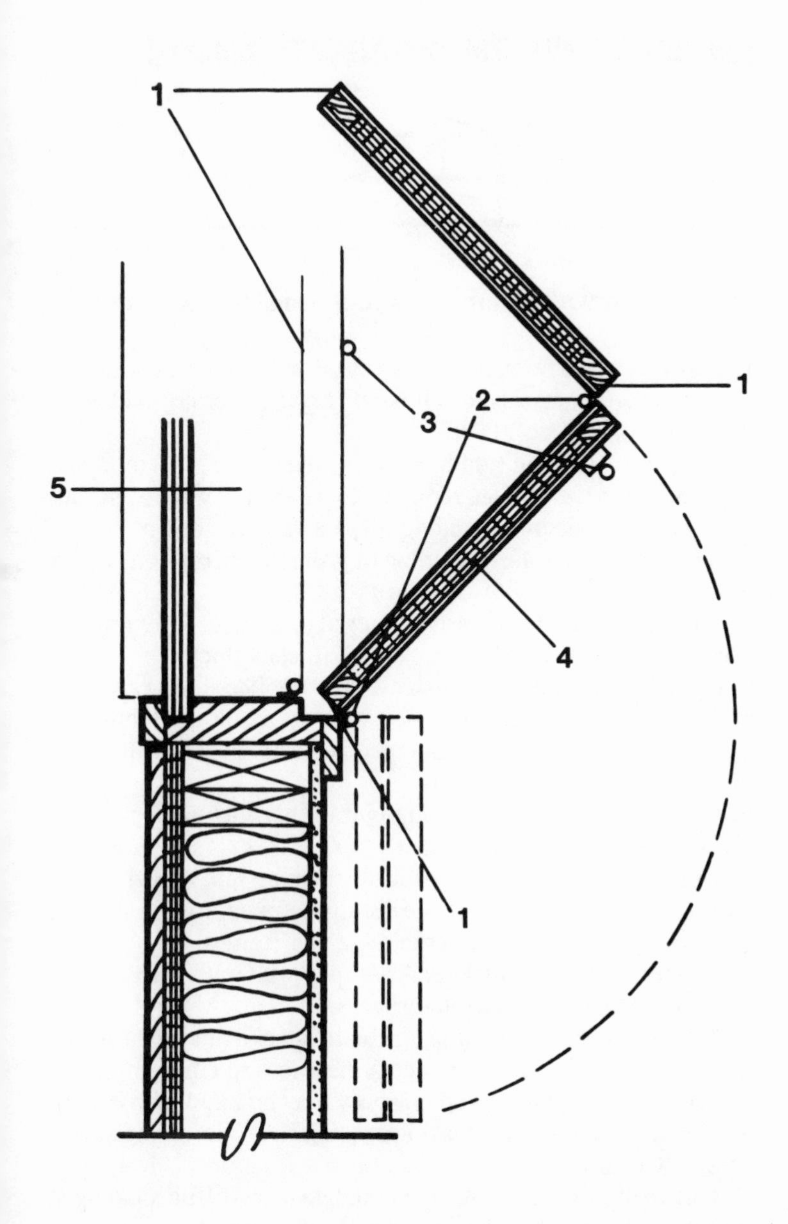
1
1
2
3
5
4
1

Radiant Trombe Wall And Summer Shading

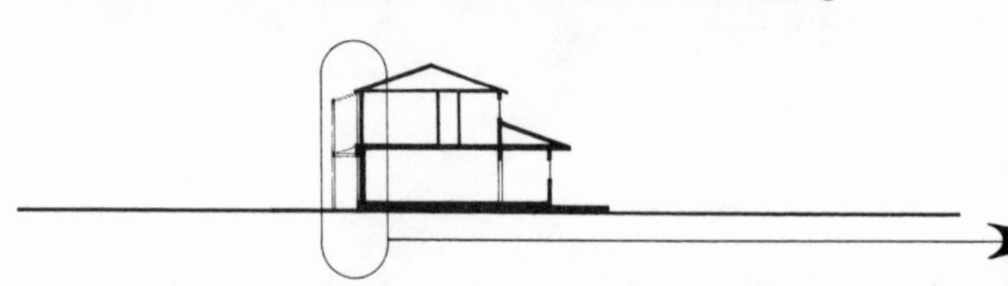

Trombe wall with summer shading, construction details

1. Control: wood or metal louvers sized and tilted to provide full shading at summer solstice (June 21) and little or no shading at winter solstice (Dec.21).
2. Collector: glazing (glass, acrylic, fiberglass, etc.) in layers appropriate to the climate mounted to be air-and water-tight but removable for cleaning; sliding glass doors are shown.
3. Absorber: dark surface color to promote absorption. Use a product capable of withstanding 150°F.
4. Storage: solid high-density material (concrete, fully grouted brick or concrete-filled block, 8 to 12 inches thick).
5. Distribution: interior finish material applied directly to storage mass with no air space or furring; book shelving and artwork hung on wall would impede effective radiant distribution of heat to adjacent room.
6. Control: movable insulating curtain permits adjustment of radiant distribution to the room.
7. Control: space between collector glazing and absorber surface should be vented to outside during summer months through top and bottom vents or operable glazing frames.
8. Conservation: thermal break at top of wall minimizes loss of heat from storage to unheated spaces.
9. Conservation: insulation on both interior and exterior of storage wall minimizes conductive heat loss to adjacent earth.
10. Conservation: unglazed masonry may be filled with insulation, if it is structurally feasible, to minimize losses of stored heat downward.
11. Control: fabric awning big enough to provide full shading at summer solstice, removable during heating season.

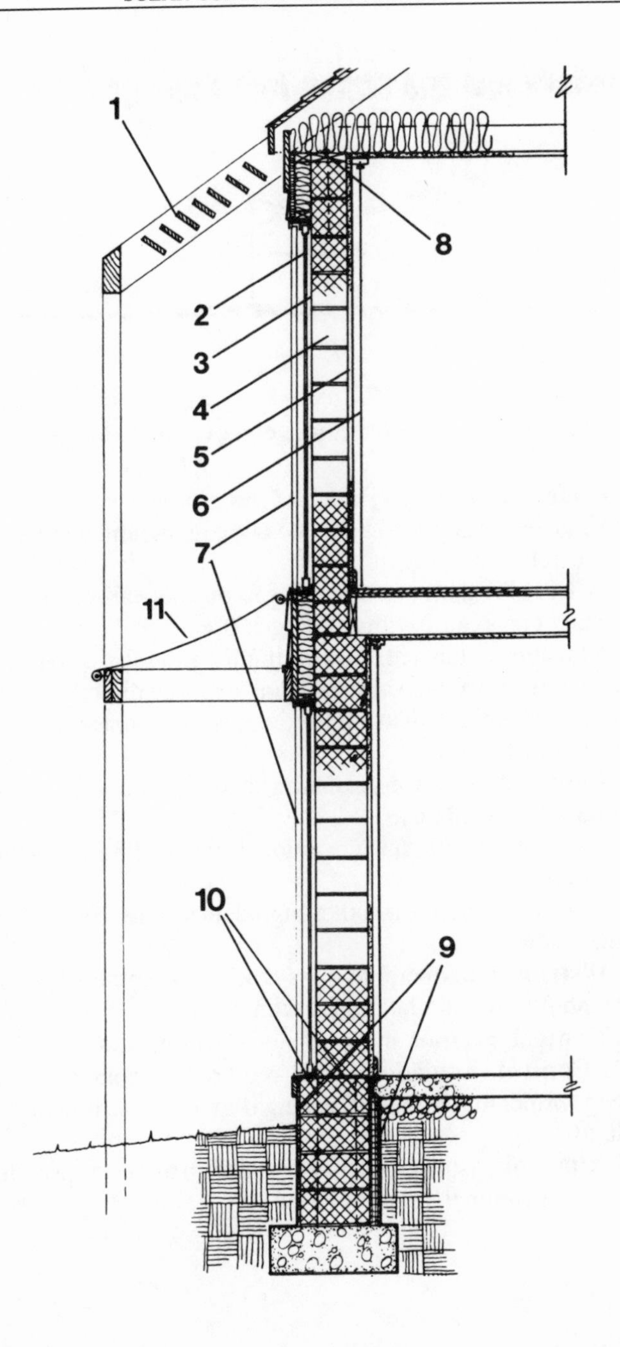
1
8
2
3
4
5
6
7
11
10
9

Greenhouse and Water Tank Storage

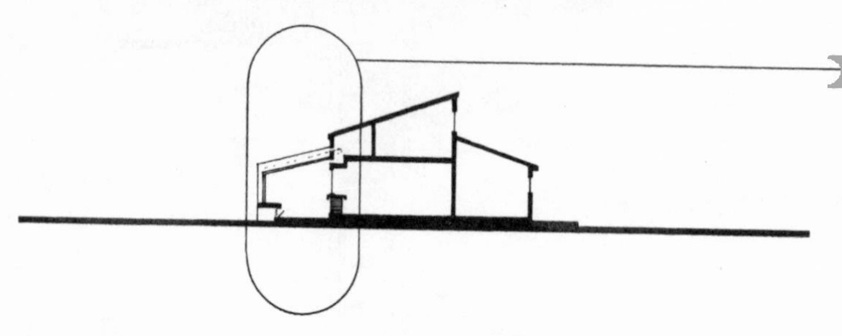

Greenhouse with water tank, construction details

1. Collector: glazing (glass, acrylic, fiberglass, etc.) in one to three layers as appropriate to the climate, mounted to be air- and water-tight.
2. Absorber: dark surface color to promote absorption. Use a product capable of withstanding 150°F.
3. Storage: water with rust inhibitor in welded metal tank.
4. Distribution: face of metal tank exposed directly to adjacent room for radiant distribution, vents in countertop aid convection distribution.
5. Control: roll-down insulation shade reduces heat loses from storage to greenhouse.
6. Absorber: dark surface color, little shading by plants and furniture.
7. Storage: concrete or other high density material with insulation below.
8. Distribution: operable glass that allows surplus hot air from greenhouse to circulate to adjacent room.
9. Control: exterior shade to block summer sun.
10. Control: insulation shade required in colder climates to keep temperatures above freezing if greenhouse is to be used for plants.
11. Control: insulated low wall vent with tight compression seal opens in summer for ventilation.

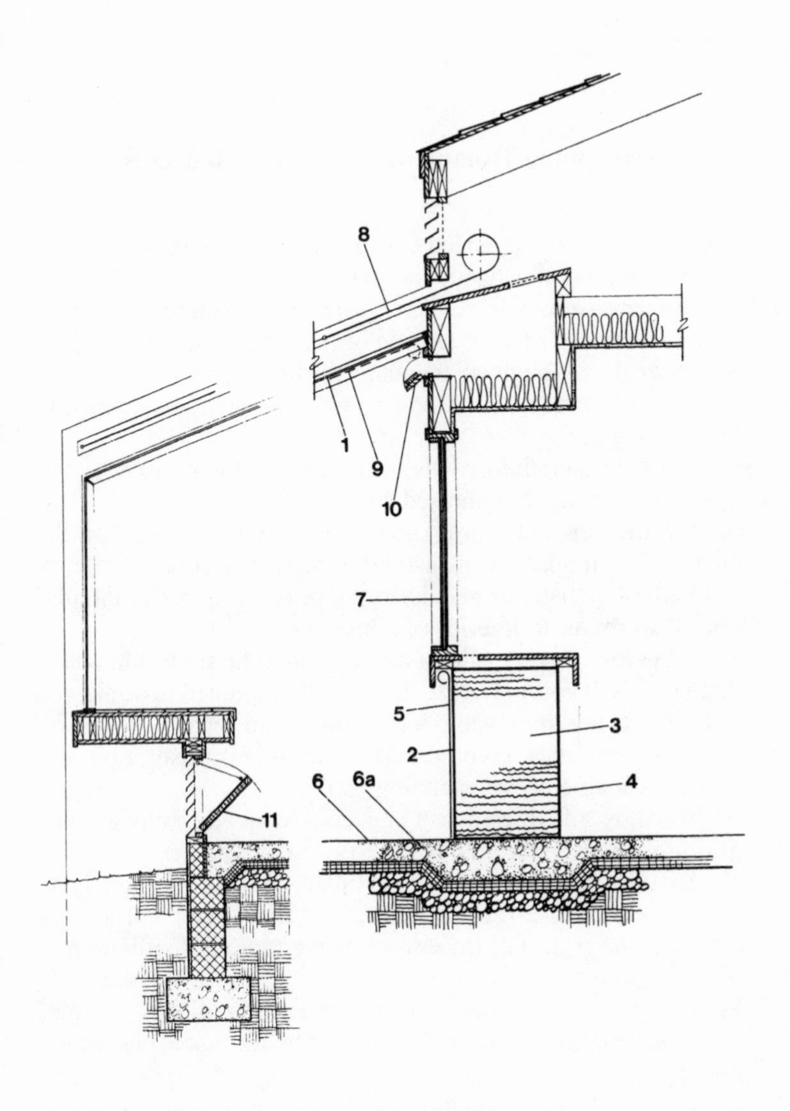
8
1
9
10
7
5
3
2
4
6
6a
11

Thermosiphoning Trombe Wall

Thermosiphoning Trombe wall construction details

1. Control: overhang, sized to provide full shading of collector glass at summer solstice (June 21).
2. Conservation: attic vents, and space between top of insulation and bottom of roof sheathing promotes attic ventilation.
3. Control: automatic or manual roll-down insulation.
4. Control: sweep minimizes the amount of heated air lost to unheated space.
5. Conservation: thermal break at top of wall minimizes loss of heat from storage to unheated space.
6. Distribution: outlet area should equal approximately half the area or the air space as measured in a cross-section.
7. Control: register closed during summer to minimize circulation of warm air to interior of house.
8. Collector: glazing (glass, acrylic, fiberglass, etc.) in one to three layers as appropriate to the climate, mounted to be air- and water-tight, and removable for cleaning and repair.
9. Absorber: dark colored surface promotes absorption; use product capable of withstanding 150°F.
10. Storage: solid high density material (concrete, fully grouted brick, or filled concrete block 8 to 12 inches thick).
11. Distribution: interior finish material applied directly to storage mass with no air space or furring; bookshelf and artwork covering the wall will impede effective radiant distribution of heat to adjacent room.
12. Control: flexible backdraft damper automatically prevents cool exterior air from settling into the room during winter nights.
13. Conservation: insulation on both interior and exterior of storage wall lessens conductive heat loss to adjacent earth.

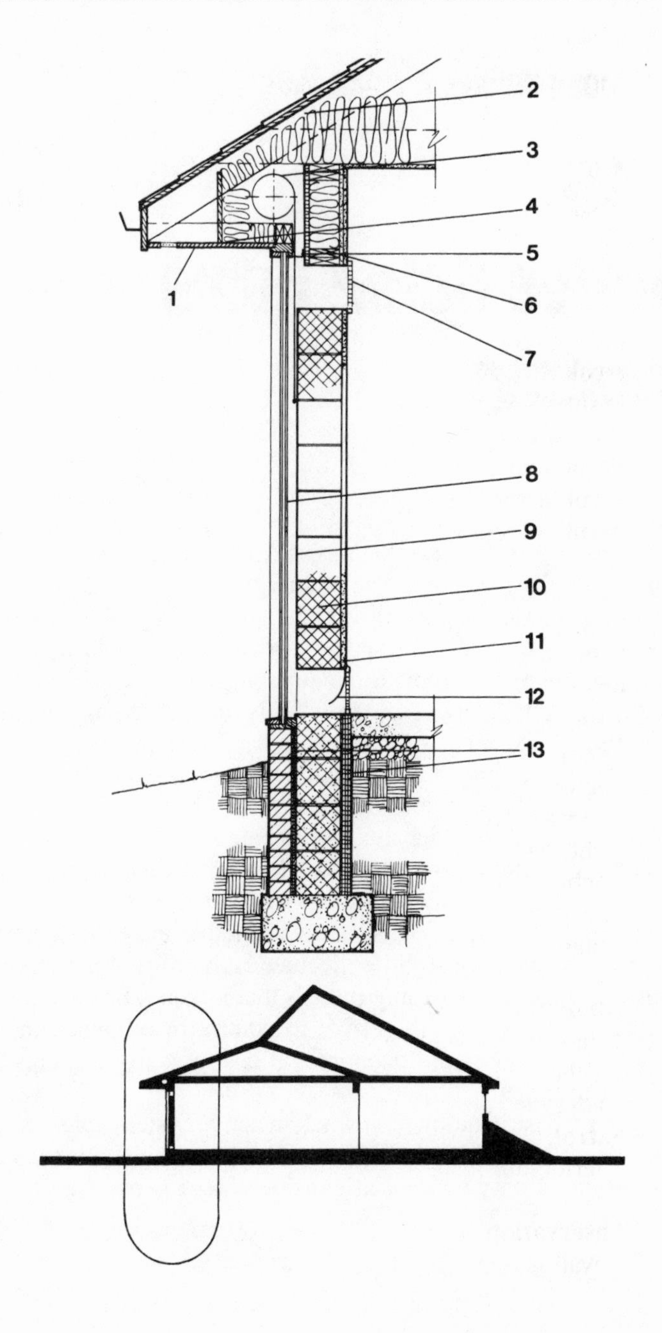
1
2
3
4
5
6
7
8
9
10
11
12
13

15.
Active Solar Home Heating

This book acknowledges that passive solar heating (see page 2) is by far the best way to use solar energy to heat a home. Active solar heating was the main thrust of solar heating up until the last few years, but active solar has proven itself to be quite a dinosaur. There are simply too many things that can go wrong in an active system and they frequently do go wrong. Active systems are also very expensive compared with passive.

So what you don't know about active solar won't hurt you, *unless* you try to heat your house with it. There is no longer an active versus passive debate. The debate is over.

The indictment: To begin with, the cost range of active solar heating systems starts where the cost range of passive systems ends. Next, while passive systems are rather easy to maintain because they involve so little mechanical hardware, active systems are subject to so many pitfalls that it would be difficult to enumerate them. And the effort to control or eliminate these pitfalls often causes the phenomenal cost of active systems to grow even larger.

Here's a basic rundown on what an active system consists of:

- Start with 200 to 500 square feet of flat-plate solar collectors to be mounted on your roof at considerable expense (see Solar Collectors, p.72);

- Then there's a large storage tank or bin to be installed in your basement, also at considerable expense. This weighty component will take up quite a bit of room and may require additional structural support to be added to the foundation;
- Follow these with an extensive system to link the pieces and to distribute heat, the full length of which must be insulated and carefully protected against leaks;
- And top it all off with an electronic control unit that can damage the system if it fails to do its job or does it the wrong way.

There's no reason why anyone should own or want to own such a system when something much simpler, much more effective, and much less expensive exists, namely passive systems. Active hardware does have a very important role in domestic water heating, where it remains simple and effective if properly installed and maintained. But for heating an entire house, forget it. Don't waste your time on active solar heating.

16. Solar Collectors

Flat Plates

The terms *solar collector* and *solar panel* most often refer to what is more accurately called the **flat-plate solar collector**. Flat-plate collectors were once considered the centerpieces of solar space heating systems.

The advent of more durable and far less expensive *passive solar heating* concepts has taken emphasis away from flat plates. They now are used more appropriately in solar hot water systems, where they operate under less complex circumstances and are less of an undertaking to install and maintain.

The typical flat-plate collector panel used in water heating is an insulated water-tight box containing a dark solar absorber plate under one or more layers of glass or transparent plastic. The dark absorber soaks up heat from the sunlight passing through the glass. Heat is transferred to a fluid flowing past or through the absorber. The heated fluid is then carried away from the collectors to a storage tank.

There are, broadly, two types of flat-plate collectors: liquid cooled and air cooled. Liquid-cooled collectors (see *Figure 1*) circulate either water or an anti-freeze solution to absorb heat. Air circulates to absorb the heat in an air-cooled collector (*Figure 2*). Liquid collectors are far more frequently used than air collectors, principally because liquid has a far greater heat absorbing capacity.

The choice of absorber plate—whether it should be copper,

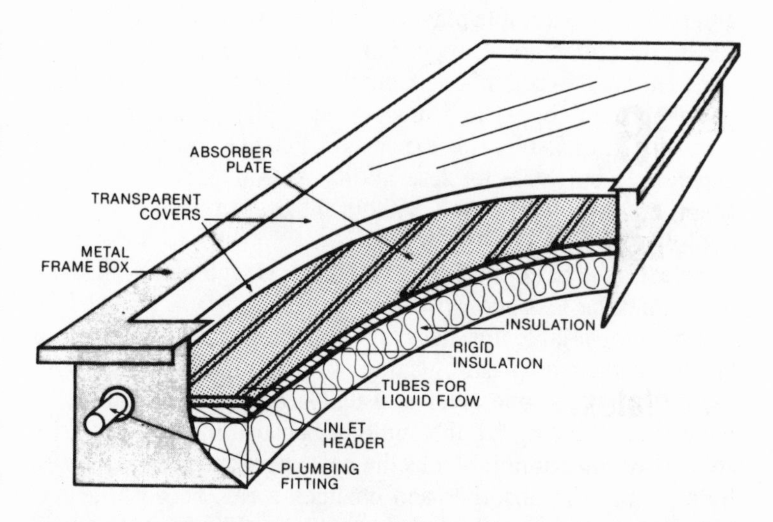

Figure 1. Flat plate collector that uses liquid to absorb and transport heat

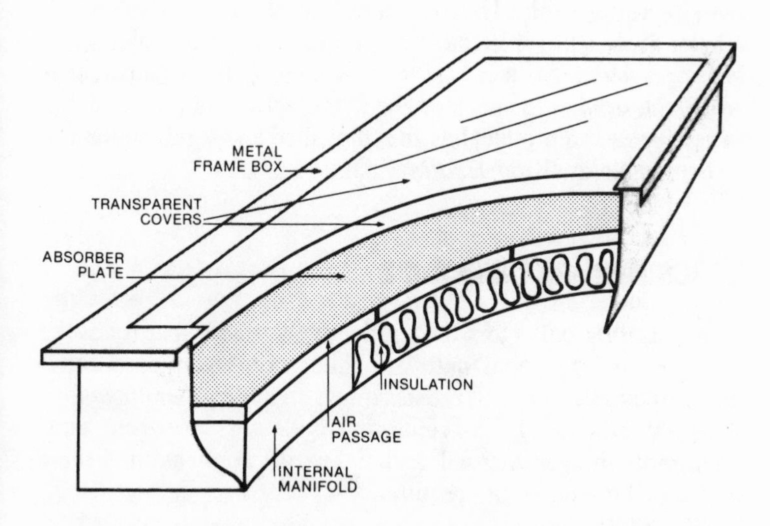

Figure 2. Flat plate collector that uses air to absorb and transport heat

aluminum, or a durable plastic—is the most important decision for liquid collectors. Copper is most expensive, yet has good durability and excellent conductivity. Aluminum conducts heat less well than copper and must be carefully protected against corrosion, but also is less expensive. Plastics are usually least expensive, but have far less ability to conduct heat. Plastic absorbers are widely used without glazing to heat swimming pools.

Selection of materials for air collectors is far less troublesome but air has far less heat absorbing capacity.

Another variable for flat-plate collectors is the number of layers of glass or plastic glazing. Colder climates usually require two layers; one layer is usually acceptable for mild climates. The glazing admits sunlight because light is primarily shortwave radiation: it blocks the heat energy that results when light strikes the absorber and changes form, because heat is longwave radiation, which does not go readily through a solid object like glazing.

There are several hundred kinds of flat-plate collectors being manufactured in the United States and abroad. Homeowners almost always buy collectors as part of a complete solar water heating system. So it is best to view them in this context. It is worth the trouble to check on the manufacturer of the collector, to verify that the product has an established and solid reputation. (Also see *Solar Water Heating*, page 75).

Concentrating Collectors

Concentrating collectors have reflective surfaces that focus the sunlight striking them onto a smaller point—an absorber—within the collector. This creates much higher temperatures than flat-plate collectors. Concentrating collectors are used most commonly in commercial and industrial applications where higher temperatures are required.

17.
Solar Water Heating

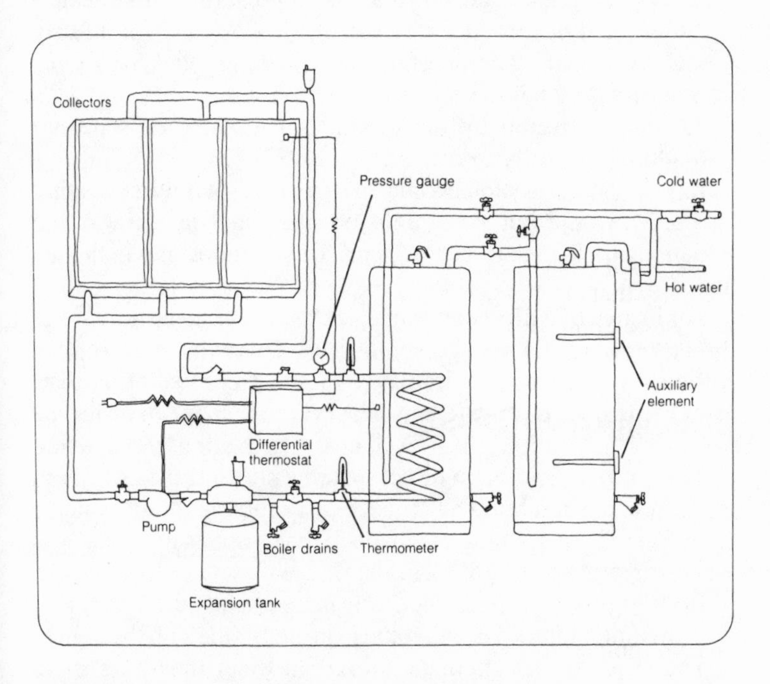

Figure 1. Schematic diagram of a typical solar water-heating system

Solar water heating is easily the most common, if not the most spectacular, use of solar energy in the United States and the world—especially in Japan, Israel, the West Indies, and several other countries.

A typical solar hot water system consists of two or three roof (or ground) mounted flat-plate solar collectors, a basement hot water tank, a pump, temperature sensors, and an electronic control mechanism that turns the system off and on.

You can purchase all of these separately, or as part of a single package that includes installation. There are good deals and bad deals. The wise buyer should be aware of a few things.

- The size of the solar collector depends on the amount of hot water needed for the household, based on the number of family members and the local climate (see *Table 1*).
- The qualifications and professional reputation of the dealer/installer and his product are critical. (See *Table 2* for a fairly extensive means of comparing dealer/installers and *Table 3* for a means by which to compare systems).
- The items covered by the product's guarantee or warranty should be carefully examined.

It is not necessary to become an expert on solar water heating before you buy, but you might be interested in some of the variations on the basic system, and the situations to which they apply.

For instance, a **direct** system circulates your household water directly through the roof-top collectors. This is in contrast to an **indirect** system which circulates an anti-freeze solution that, after being heated, transfers its heat (via a mechanism inside the hot water tank called a heat-exchanger) to the household water.

The anti-freeze used in direct systems prevents the collectors from freezing and bursting. Freeze protection in direct systems is provided by a draining mechanism that empties the collectors when there is no sun to warm them.

Both approaches have their advantages and disadvantages, with manufacturers taking stands on both sides of the fence.

The systems that circulate household water may have more problems with corrosion. On the other hand, indirect, anti-freeze systems are less efficient because they use heat-ex-

Note: The collector area indicated in each case can be expected to provide, at the lowest possible cost per unit of heat produced, at least 50 percent of the hot water used annually by the average household of that size. A collector panel typically has 20 square feet of collecting surface.

Location	Number of Users 2	4	6	Location	Number of Users 2	4	6
	Square Feet				Square Feet		
BIRMINGHAM, AL	40	60	80	BILLINGS, MT	40	60	80
FAIRBANKS, AK	60	80	100	GREAT FALLS, MT	40	60	80
TUCSON, AZ	40	60	60	LINCOLN, NE	40	60	80
LITTLE ROCK, AR	40	60	80	LAS VEGAS, NV	40	60	60
LOS ANGELES, CA	40	60	80	RENO, NV	40	60	80
SACRAMENTO, CA	40	60	60	CONCORD, NH	60	80	100
SAN FRANCISCO, CA	40	60	80	ATLANTIC CITY, NJ	40	60	80
DENVER, CO	40	60	80	ALBUQUERQUE, NM	40	60	60
GRAND JUNCTION, CO	40	60	80	ALBANY, NY	60	80	100
HARTFORD, CT	60	80	100	NEW YORK, NY	60	80	100
WILMINGTON, DE	60	80	100	ROCHESTER, NY	60	80	100
WASHINGTON, DC	60	80	100	SYRACUSE, NY	60	80	100
JACKSONVILLE, FL	40	60	80	CAPE HATTERAS, NC	40	60	80
MIAMI, FL	40	60	80	RALEIGH, NC	40	60	80
TALLAHASSEE, FL	40	60	80	BISMARCK, ND	40	80	100
TAMPA, FL	40	60	80	CLEVELAND, OH	60	80	100
ATLANTA, GA	40	60	80	COLUMBUS, OH	60	80	100
SAVANNAH, GA	40	60	80	OKLAHOMA CITY, OK	40	60	80
HILO, HI	40	60	80	TULSA, OK	40	60	80
HONOLULU, HI	40	60	60	MEDFORD, OR	40	60	80
BOISE, ID	40	60	80	PORTLAND, OR	60	80	100
POCATELLO, ID	40	60	80	PHILADELPHIA, PA	60	80	100
CHICAGO, IL	60	80	100	PITTSBURGH, PA	40	80	80
PEORIA, IL	40	60	80	STATE COLLEGE, PA	60	80	100
INDIANAPOLIS, IN	60	80	100	PROVIDENCE, RI	60	80	100
DES MOINES, IA	40	80	80	CHARLESTON, SC	40	60	80
WICHITA, KS	40	60	80	RAPID CITY, SD	40	60	80
LEXINGTON, KY	60	80	100	NASHVILLE, TN	40	60	80
LOUISVILLE, KY	60	80	100	AMARILLO, TX	40	60	80
NEW ORLEANS, LA	40	60	80	DALLAS, TX	40	60	80
SHREVEPORT, LA	40	60	80	EL PASO, TX	40	60	60
CARIBOU, ME	60	80	100	HOUSTON, TX	40	60	80
PORTLAND, ME	60	80	100	SALT LAKE CITY, UT	40	60	80
BALTIMORE, MD	60	80	100	BURLINGTON, VT	60	80	100
AMHERST, MA	60	80	100	MT. WEATHER, VA	60	80	80
BOSTON, MA	60	80	100	NORFOLK, VA	40	60	80
LANSING, MI	60	80	100	RICHMOND, VA	60	80	80
SAULT STE. MARIE, MI	60	80	100	SEATTLE, WA	60	80	100
MINN.-ST. PAUL, MN	60	80	100	SPOKANE, WA	40	60	80
JACKSON, MS	40	60	80	CHARLESTON, WV	60	80	100
KANSAS CITY, MO	40	60	80	MADISON, WI	60	80	100
ST. LOUIS, MO	40	60	80	CASPER, WY	40	60	80
				CHEYENNE, WY	40	60	80

Table 1. Collector sizes suggested to provide solar hot water for a family of four, in selected U.S. cities

changers, and the anti freeze must be checked often for chemical breakdown, and periodically replaced. Anti-freeze is toxic, so steps must be taken to isolate it from any possible connection with the household water supply. These are all problems that can be successfully solved, however.

These types of considerations reinforce the need for finding a dealer/installer who has really mastered either or both of these systems. Both types work, but they need to be planned and installed by someone who really understands what makes a system succeed.

Making Sure

Proven Workability—Before you sign a contract with a dealer/installer it's very important to find out if the system he proposes to install has a good track record. Ask for references from other customers with working systems. Contact these people. Ask how long their systems have been working, if there have been problems, if the dealer has been prompt and cooperative in correcting problems, and whether or not they are satisfied with the system as a whole. The approval of a homeowner with the same system is a very good sign.

Maintenance—When checking references, compare what system owners say about the need for maintenance with what the dealer has told you. Find out how much effort and expense are needed to keep the system operating well.

Warranties—A warranty of three to five years on major system components is a reasonable expectation. Although it is more important to have a good system than a good warranty, it is also reasonable to expect that a good system *will* have a good warranty. Read the warranty carefully, making sure that it is comprehensive and that the dealer/installer is directly responsible to you for warranty coverage. Since some system parts are likely to come from different manufacturers, make sure that the installer's warranty not only covers defects resulting from improper installation but also covers all of the equipment.

Contracts—Read the contract carefully. Everything should be spelled out in writing at a price that is incontestably *firm*.

The contract should include such items as a detailed description of the work to be done, the total price, approximate dates for the beginning and completion of work, and a general description of the warranty to be provided, in addition to the customary certifications about liability, liens, etc.

You may also want to limit the interruption of your hot water service (one day should suffice), and stipulate that a percentage of the contract price will be withheld pending re-inspection by the installer after the first few weeks of operation. It is also wise to spell out the installer's responsiblity to restore disturbed roofing, siding, building insulation, walls, and landscaping upon completion.

Installation—The installation process should go smoothly if done by someone qualified. Make sure that the collectors are oriented to face within 30 degrees of true south and that they are tilted at an angle close to the latitude of your location. The installer should bring the system up to full operation for an initial shakedown and should return within the next month to see if it is working properly.

Finding A Dealer/Installer

A solid word-of-mouth recommendation from a person whose judgment you trust is always the best way to find anyone who is selling you his services. Or you can simply look in the Yellow Pages either under solar energy or under plumbing and heating contractors for dealer/installers of solar water heating systems.

Another way to find a dealer/installer is to look up the name of a manufacturer of systems from the list that follows and call or write to them for the name of a certified dealer in your area. Your choice of manufacturer could be based either on how close they are to you or on your recognition of a reputable nationwide corporation.

Still another way to locate a dealer is to consult the *Sources of Information* section of this book under your state for an organization that could give you the names, or possibly offer a recommendation for dealers near you.

Remember that up to 40 percent of the cost of a solar water heating system can be taken as a direct federal income tax credit. *Table 4* indicates how much is deducted from what you owe the federal government on systems costing from $1,000 to $3,500.

QUALIFICATIONS	Individuals or Firms A	 B	 C
	(√) indicates description applies to individual or firm		
Solar installation experience			
Satisfied customers			
Solar-related background			
Nearby store or office			
Stable, established business			
Regular business hours			
Assistance in obtaining financing			
Good complaint response record			
Strong manufacturer support			
Willingness to negotiate terms			
Good warranty terms offered			
On-site repairs made			
Professional standing			
Familiarity with building regulations			
Willingness to give full explanations			
Ongoing service offered			
Frequently used parts in stock			

Table 2. Checklist for comparing solar dealers/installers

	Systems A	B	C
COST—including installation			
	(√) indicates description applies to system		
BENEFITS			
Efficiency			
Proven workability			
Collector test results available			
Working system available to inspect			
Durability			
Well-made collectors			
Reputable brands			
Low maintenance requirements			
Easily accessible components			
Good protection against freezing			
Good protection against corrosion			
Good protection against leaks			
Built-in safety precautions			
Manufacturer's Commitment			
Installer training provided			
Replacement parts stocked			
Repairs made on site or costs covered			
Good warranty offered through dealer			
Complete owner's information provided			
Monitoring Devices			
Thermometers and gauges included			
Pump operating light included			
Devices available as options			

Table 3. Checklist for comparing solar water-heating systems

Note: To find the credit on values not appearing in the table, follow this simple method:

If the installed cost is $10,000 or less, multiply it by 40 percent.

If Your Installed Solar Hot Water System Cost Is: $	Then Your Tax Credit Is: $
1000	400
1250	500
1450	580
1600	640
1750	700
1900	760
2050	820
2300	920
2550	1020
2750	1100
3000	1200
3200	1280
3500	1400

Federal tax credits for solar water-heating systems

18.
Some Manufacturers of Domestic Hot Water Systems

Liquid Systems

Alabama

Solar Unlimited
37 Traylor Island
Huntsville, AL 35801
(205) 534-0661
Trade name:Suncatcher

Aircraftsman
PO Box 628
Millbrook, AL 36054

Arizona

Sunshine Unlimited
900 North Jay St.
Chandler, AZ 85224
(602) 963-3878

Solar Energy Applications, Inc
1102 E. Washington St.
Phoenix, AZ 85034
(602) 244-1822

Goettle Air Conditioning
2005 E. Indian School Rd.
Phoenix, AZ 85016
(602) 957-9800

Copper State Solar Products, Inc
4610 S. 35th St
POB 20504
Phoenix, AZ 85036
(602) 276-4221

Sunpower Systems Corp
510 S 52nd St
Suite 101
Tempe, AZ 85281
(602) 894-9671

Sun-West Solar Systems Inc
Sunland Inc
1422 W. 23rd St/Suite 105
Tempe, AZ 85282
(602) 894-9671
Trade name: Suntrap

Cornell Energy, Inc
245 S Plumer
Suite 29
Tucson, AZ 85719
(602) 882-4060

California

Piper Hydro, Inc
3031 E. Coronado
Anaheim, CA 92806
(714) 630-4040

Ra-Los, Inc
559 Union Ave.
Campbell, CA 95008
(408) 371-1734

Solar King International Inc
8577 Canoga Ave.
Canoga Park, CA 91304
(213) 998-6400

Solargenics Inc
9831 Mason Ave.
Chatsworth, CA 91311
(213) 998-0806

Helix Solar Systems
245 S. Eighth Ave
City of Industry, CA 91744
(213) 961-0471

Solpower
10211-C Bubb Rd
Cupertino, CA 95014
(408) 996-3222
Trade name: Solpower

Coppersmiths
8892 Walker Ave.
Cypress, CA 90630
(714) 761-2758

Mr. Sun Inc
1019 S Main St
Fallbrook, CA 92028
(714) 728-0553
Trade name: Mr. Sun

Ying Manufacturing Corp
1957 W. 144 St.
Gardena, CA 90249
(213) 770-1756

Sav Solar Systems, Inc
550 W Patrice Pl. Suite A
Gardena, CA 90248
(213) 327-7210
Trade name: Sav

EPI
1424 W 259th St
Harbor City, CA 90710
(213) 539-8590
Trade name: Sun Wizard

Intersun, Inc
PO Box 907
7271 Murdy Circle
Huntington Beach, CA 92647
(714) 848-8188

Solar Energies of California
11421 Woodside Ave. South
Lakeside, CA 92040
(714) 448-4300

Advanced Energy Technology, Inc
121-C Albright Way
Los Gatos, CA 95030
(408)866-7686

Environmental Energy Management System
Carmel Hill Professional Center
23845 Hollman
Monterey, CA 93940
(408) 625-3364
Trade name: Solatherm

Alten Corp
2594 Leghorn St.
Mountain View, CA 94043

(415) 969-6474
Solar Pacific, Inc
11145 Vanowen St
North Hollywood, CA 91605
(213) 877-2678

Energy Specialties
9312 Greenback La.
Orangevale, CA 95662
(916) 988-1208

Western Energy Inc
2652 E Bayshore Rd.
Palo Alto, CA 94303
(415) 237-9614

Heliodyne Inc
770 S 16th
Richmond, CA 94804
(415) 237-9614

Sunburst Solar Energy Inc
4131 A Power Inn Rd
Sacramento, CA 95826
(916) 739-8485

Kaiser Energy Engineering
975 Terminal Way
San Carlos, CA 94070
(415) 593-1463
Trade name: Keesol

Energy Systems, Inc
4570 Alvarado Canyon Rd.
Building D
San Diego, CA 92120
(714) 280-6660
Trade name: Esi-Save

Sunspot Environmental Energy Systems
3007 N. Euclid
San Diego, CA 92105
(714) 452-0252

Solahart International, Inc
3560 Dunhill St.
San Diego, CA 92121
(714) 452-0252

Energy Harvester
11807 Bernardo Ter
San Diego, CA 92128
(714) 452-0252
Trade name: SPP-4

Technitrek Corp.
1999 Pike Ave.
San Leandro, CA 94577
(415) 352-0535

American Appliances Corp
2341 Michigan Ave
Santa Monica, CA 90404
(213) 870-8541
Trade name: Solarsteam
Direct-Flo

Environmental Energy Management & Manufacturing Corp
2722 Temple Ave.
Signal Hill, CA 90806
(213) 427-0991
Trade name: Sunsorber

Buckmaster Industries
PO Box 1855
Sunnymead CA 92388
(714) 653-8461

Colorado

Entropy Ltd
5735 Arapahoe Ave.
Boulder, CO 80303
(303) 443-5103
Trade name: Sunpump
Suncycle

American Heliothermal Corp
Galleria S. Tower Suite 450
Denver, CO 80222
(303) 753-0921
Trade name: The Solar Appliance

R-M Products
5010 Cook St
Denver, CO 80216
(303) 825-0203

Solar Specialties Inc
Rte. 7, Box 409
Golden, CO 80401
(303) 642-3063

Solar Energy Research Corporation
10075 East County Line Rd
Longmont, CO 80501

Connecticut

Solar Products Manufacturing
Alcap Ridge
Cromwell, CT 06416
(203) 635-0267
Trade name: Sunspot

American Solar Heat Corp
7 National Pl
Danbury, CT 06810
(203) 748-5554

Solar Processes Inc
11 Velvet Lane
Mystic, CT 06353
(203) 536-0430

Solar Craft Industries
45 Hayden Station Rd.
Windsor, CT 06095
(203) 688-7383

Delaware

Forter Energy Products
PO Box 827
Newark, DE 1971
(301) 398-0284
Trade name: Solaris

Solar Energetics, Inc
301 S. West St.
Wilmington, DE 19801
(302) 654-3252
Trade name: Sunpower

Florida

OEM Products Inc
Solarmatic
Rt 3, Box 295
Dover, FL 33527
(813) 752-3121
Trade name: Solarmatic

Semco Corp
5701 NE 14th Ave
Ft. Lauderdale, FL 33334
(305) 776-1300

Solar Energy Products Inc
PO Box 1048
Gainesville, FL 32601
(904) 377-6527
Trade name: Sunfired

US Solar Corp
PO Drawer K
Hampton, FL 32044
(904) 468-1517
Trade name: Eagle Sun

Wilkes Sun Energy Systems
PO Box 1842
Hobe Sound, FL 33455
(305) 546-3139

D W Browing Contracting Co
475 Carswell Avenue
Holly Hill, FL 32017
(904) 252-1528

Suntrak Inc
Solar Industries of Florida
3231 Trout River Blvd
Jacksonville, FL 32208
(904) 768-4323
Trade name: Suntrak

Solar Heater Manufacturer
1011 6th Ave South
Lake Worth, FL 33460
(305) 586-3839

Energy Conservation Equipment Corp
1527 C Road
Loxahatchee, FL 33470
(305) 793-0851
Trade name: Ece Eagle

Solar Products
Sun Tank Inc
4291 Northwest 7th Ave
Miami, FL 33127
(305) 756-7609

W.R. Robbins and Son
1401 NW 20th St
Miami, FL 33142
(305) 325-0880
Trade name: Floridian

American Sun Corporation
9300 S. Dadeland Blvd. #703
Miami, FL 33156
(305) 681-2501

American Sunsystems Inc
8454 NW 58th St
Miami, FL 33166
(305) 592-1361

Solar Water Heaters of New Port Richey
1214 US Highway 19 North
New Port Richey, FL 33552
(813) 848-2343

Solar Energy Systems, Inc
1065 NE 125th St.
North Miami, FL 33161
(305) 893-44420

Sun Harvesters Inc
416 NE Osceola
Ocala, FL 32670
(904) 629-0687

Heliokon Industries Inc
138 Industrial Loop West
Orange Park, FL 32073
(904) 264-6453

Largo Solar Systems Inc.
991 SW 40th Ave
Plantation, FL 33317
(305) 583-8090

Solar Development Inc
3630 Reese Ave
Garden Industrial Park
Riviera Beach, FL 33404
(305) 822-3689

Energy Transfer Systems
5001 W Waters
Tampa, FL 33614
(800) 237-5057

Georgia

Rheem Manufacturing Co
5780 Peachtree-Dunwoody Rd
Atlanta GA 30342
(404) 256-2037
Trade name: Sun Set

National Solar Supply
2331 Adams Dr NW
Atlanta, GA 30318
(404)352-3478

United States Solar Industries
5600 Roswell Rd
Prada E., Ste. 350
Atlanta, GA 30342
(404) 252-1870
Trade name: Sun Rise

Energy Converters Inc
RD 1 Johnson Rd. Box 92
Chickamauga, GA 30707
(615) 624-1594

Ali
5965 Peachtree Corners East
Norcross, GA 30071
(404) 449-5900

Iowa

Engineers Ltd
823 Central
Dubuque, IA 52001
(319) 556-7544

Illinois

A O Smith Corp
Box 28
Kankakee, IL 60901
(815) 933-8241
Trade name: Conservationist

Kansas

Alternate Energy Sources Inc
752 Duvall
Salina, KS 67401
(913) 825-8218
Trade name: Sun Grabber®

Massachusetts

Megatech Corp
29 Cook St.
Billerica, MA 01866
(617) 273-1900
Trade name: Megatech

Acorn Structures Inc
PO Box 250
Concord, MA 01742
(617) 369-4111
Trade name: Sunwave

Dixon Energy Systems Inc
7 East St
Hadley, MA 01035
(413) 584-8831
Trade name: Drainback

Columbia Chase Solar Energy Div.
55 High St.
Holbrook, MA 02343
(617) 767-0513

Elbart Manufacturing Co
127 West Main St.
Millbury, MA 01527
(617) 865-9412

Sun Systems Inc
POB 347
Milton, MA 02186
(617) 265-9600

Solar Aqua Heater Corp
15 Idlewell St.
Weymouth, MA 02188
(617) 843-7255

Maryland

Solar Energy Systems & Products
500 N. Alley
Emmitsburg, MD 21727
(301) 477-6355

Futuristic Solar Systems
4900 Beech Pl.
Temple Hills, MD 20031
(301) 899-3430

Maine

Dumont Industries
Main St.
Monmouth, ME 04259
(207) 933-4811

Michigan

Refrigeration Research Solar Research
525 N. 5th St.
Brighton, MI 48116
(313) 227-1151

Minnesota

Solargizer International Inc
2000 West 98th St.
Bloomington, MN 55431
(612) 888-0018

Solergy Co
7216 Boone Ave. N
Minneapolis, MN 55428

Missouri

Tri Development Corp
500 NE 23rd St.
PO Box 4529
Rolla, MO 65401
(314) 341-3100
Trade name: Solarray

Weather-Made Systems Inc
West Hwy 266, Rt 7, Box 300-D
Springfield, MO 65802
(417) 865-0684

Montana

Sunset Solar Construction
3 NE 915 Pine Hollow Rd.
Sevensville, MT 59870
(406) 777-3168

North Carolina

Hardy & Newson Inc
PO Box 158
La Grange, NC 28551
(919) 566-3111
Trade name: Sunmate

New Jersey

Solar Thermal Systems
PO Box 592
Florham Park, NJ 07932
(201) 765-4200
Trade name: Daystar

Solar Living Inc
PO Box 12
Netcong, NJ 07857
(201) 691-8483

Sunworks
Div of Sun Solector Corp.
POB 3900
Somerville, NJ 08876
(201) 469-0399
Trade name: Solector

Solar and Geophysical Engineering Inc
PO Box 576
Sparta, NJ 07871
(201) 383-9230

Creighton Solar Concepts
Woodbine Airport
Woodbine, NJ 08270
(609) 861-2442

New Mexico

Zomeworks Industries
PO Box 712
Albuquerque, NM 87103
(505) 242-5354

New York

Advanced Energy Technologies
Solartown USA
Clifton Park, NJ 12065
(518) 371-2140
Trade name: Solar Homeside

Prima Industries Inc
1719 Great Neck Rd.
Copiague, NY 11726
(516) 691-5600

Sunmaster Corp
12 Spruce St.
Corning, NY 14830
(607) 937-5441
Trade name: Sunmaster

Solar Dynamics Systems Corp
3230 N. Jasper Rd.
Jasper, NY 14855
(607) 792-4722

Amcor Group Ltd
350 5th Ave.
Suite 1907
New York, NY 10001
(212) 736-7711

Nortec Solar Industries Inc
PO Box 698
Ogdensburg, NY 13669
(315) 425-1255

Revere Solar And Architectural Products, Inc
PO Box 151
Rome, NY 13440
(315) 338-2401
Trade name: Sun-Pride

Grumman Energy Systems Inc
4175 Veterans Memorial Hgwy
Ronkonkoma, NY 11779
(516) 244-2700
Trade name: Sunstream

Environment/One Corp
2773 Balltown Rd.
Schnectady, NY 12309
(518) 346-6161

Ohio

Alpha Solarco
1014 Vine St., Suite 2230
Cincinnati, OH 45202
(513) 621-1243

Mor-Flo Industries Inc
18450 South Miles Rd.
Cleveland, OH 44128
(216) 663-7300
Trade name: Solarstream

Solar Central
7213 Ridge Road
Mechanicsburg, OH 43044
(513) 828-1350

Oklahoma

Hefron Solar
501 N. Meridian, Suite 403
Oklahoma City, OK 73107
(405) 947-5551

Saves/Naturgy Inc
533 S. Rockford
Tulsa, OK 74120
(918) 587-7176
Trade name: Naturgy

Oregon

Scientifico
35985 Row River Rd.
Cottage Grove, OR 97424

Pennsylvania

Sunearth Solar Products Corp
352 Godshall Dr.
Harleysville, PA 19438
(215) 256-6648

Heliotherm Inc
W. Lenni Rd.
Lenni, PA 19052
(215) 459-9030

Amicks Solar Heating
375 Aspen St.
Middleton, PA 17057
(717) 944-4544

Puerto Rico

Solar Devices Inc
GPO BOH 3727
San Juan, PR 00936
(809) 783-1775

Rhode Island

Vulcan Solar Industries Inc
6 Industrial Dr.
Smithfield, RI 02917
(401) 231-4422
Trade name: Sunline

Tennessee

State Industries Inc
Cumberland Street
Ashland City, TN 37015
(615) 792-4371
Trade name: Solarcraft

W L Jackson Manufacturing Co
1200-26 E. 40th St.
PO Box 11168
Chattanooga, TN 37401
(615) 867-4700
Trade name: Solar Saver

Texas

Cole Solar Systems, Inc
440A East St. Elmo Rd.
Austin, TX 78743
(512) 444-2565

Lennox Industries Inc
POB 400450
Dallas, TX 75240
(214) 783-5000
Trade name: Solarmate

Yazaki Corp
Dallas Liaison Office
Dallas, TX 75234
(214) 783-5000
Trade name: Sola-Ace

Northrup Inc
302 Nichols Dr.
Hutchins, TX 75141
(214) 225-7351
Trade name: Solar Flair

American Solar King Corp
7200 Imperial Drive
Waco, TX 76710
(817) 776-3860
Trade name: Decade 80

Virginia

Sunstar II
2120 Angus Rd.
Charlottesville, VA 22901
(804) 977-3719
Trade name: Sunflow

Solar American Co. Inc
13629 Warwick Blvd.
Newport News, VA 23602
(804) 874-0836

Reynolds Metals Company
POB 27003
Richmond, VA 23261
(804) 281-4421

Virginia Solar Components Inc
Route 3, Highway 29 South
Rustburg, VA 24588
(804) 821-9523

Solar One Ltd
100 Parker Lane
Virginia Beach, VA 23454
(804) 340-7774

Vermont

Solar Alternative Inc
2 S. Main
Brattleboro, VT 05301
(802) 254-6668
Trade name: Solite

Earth Services, Inc
Box 99/Rt 30
Pawlet, VT 05761
(802) 325-3323

Washington

E and K Service Co
16824 74th Ave, NE
Bothell, WA 98011
(206) 481-4874

Air Systems

Colorado

Solar Developments Inc
3323 Moline St.
Aurora, CO 80010
(303) 343-8154
Trade name: Solstar

Solaron Corporation
300 Galleria Tower
720 S Colorado
Denver, CO 80222
(303) 759-0101

Indiana

Solar Shelter Engineering Co
800 S. Council
Muncie, IN 47302
(317) 282-1189

Massachusetts

Solafern Ltd
536 MacArthur Blvd.
Bourne, MA 02532

Nebraska

Valmont Energy Systems Inc
Valley, NE 68064
(402) 359-2201

New Hampshire

Granite State Solar Industries
PO Box 951
Dover, NH 03820
(603) 742-7685
Trade name: Airhair

New York

The Energy Outlet Inc
2 West Main St.
McGraw, NY 13101
(607) 836-6971

Ohio

Rom-Aire Solar Corp
121 Millet Road
Avon Lake, OH 44012
(216) 933-5000
Trade name: Rom-Aire

Solar 1
Division of Stellar Industries
7265 Commerce Drive
Mentor, OH 44060
(216) 951-6363

Tennessee

Production Plastics Inc
315 Garfield St.
McMinnville, TN 37110
(615) 473-4611
Trade name: Sun Duster

Texas

Alternative Energy Resources
1155-K Larry Mahan Dr.
El Paso, Tx 79925
(915) 593-1927

Wisconsin

Sun Stone Solar Energy Equipment
PO Box 138
Baraboo, WI 53913
(608) 356-7744

19. Federal Tax Credits for Solar Energy

Purchasing solar energy equipment, or wind and goethermal energy equipment, makes the purchaser eligible for a federal income tax credit. The credit allowed is 40 percent of the first $10,000 spent on materials and labor.

Active and passive systems both qualify for the credit, but for passive the credit is not allowed on any component that serves a function other than the collection, storage, distribution or control of solar energy. In other words, the typical materials and structures involved in passive solar, such as floors, support walls, and greenhouses, are not eligible. But the containers (which serve no structural function) used in water walls, for example, would be eligible.

Solar water heating systems are most certainly eligible.

The credit may be claimed by homeowners, tenants, joint owners, and occupants, and members of condominium management associations and cooperative housing corporations. The basic requirement is that the materials or systems be installed at the taxpayer's principal residence, whether it is a single-family dwelling or a unit in a multi-family building. Summer homes or vacation homes do not qualify.

All of the credit does not have to be taken in one year, but the maximum expenditure on which the credit is allowed is the $10,000. So if $3,000 was spent on a solar water heater in one year, the taxpayer may not calculate a credit on more than $7,000 in subsequent years.

Internal Revenue Service Publication 903, *Energy Credits for Individuals*, contains more details. The proper IRS form for claiming the credit is 5695: "Energy Credits."

20.
Solar Energy's Growth and Potential

Energy has been ours for the asking. As a result the United States consumes more energy than all of Western Europe. We use more energy just for air conditioning than the 900 million people of China use for all their needs.

Before 1973, Americans never gave the subject of energy much thought. Electricity, our most prized tool, was there for us at the flick of a switch. Who knew or cared how it was generated?

The use of oil, like coal and wood before it, changed the face of America. Priced at a a few dollars a barrel, oil fueled our exodus to the suburbs. It permitted the most massive road building program ever undertaken. It powers over 150 million cars, trucks, and buses (almost half of the world's vehicles).

In 1970, U.S. oil production peaked. From then on output declined, while our appetite for it continued to grow. Spot shortages developed. In 1973, import quotas were dropped so that America could buy all the oil it needed from foreign producers.

In late 1973, in response to our support for Israel during the October War, Arab oil merchants drastically reduced shipments to the United States. Then they began a long series of price increases. By December 1974, oil was eight times the price it had been just five years before.

Hooked on oil and caught without a worthy substitute, we ravaged our lands seeking more. Four times as many oil and gas wells have now been drilled in the United States as in all the rest of the world. Yet our reserves have been decreasing.

We have about 500 barrels of oil left per person. If the average American drove a car getting 10 miles to the gallon, 10,000 miles per year, our remaining reserves would last another twelve years. Fate has given this generation the task of harnessing new energy sources.

Enter *Solar Energy*. Using the sun's energy for heating is hardly a new idea. The Romans were so concerned about solar energy that in 500 AD one citizen was arrested for intruding on another's access to sunlight. Solar heating has been used for a long time in this country also. A few "solar" houses in colonial New England concentrated windows on the south side and used sloping roofs to reduce the escape of heat and to deflect winds.

Solar hot water heaters ushered in America's first Solar Age. They were used by 80 percent of the houses built in Miami between 1937 and 1941, when solar energy was cheaper than electricity, its competitor. Solar water heaters paid for themselves in just two years.

Economy-minded people demanded that developers install solar water heaters. If a house did not have one they wouldn't buy it. Selling a non-solar house in Florida was difficult. Even after the solar heaters became twice as expensive as electric heaters to install, they remained popular. Consumers knew how much they reduced utility bills.

The Second World War was a turning point for the emerging solar industry. Copper, needed for most solar collectors, could not be used for non-military purposes. Solar systems designed for small families could not keep up with the demands of the post-war baby boom. And electric rates dropped. As a result, few people bought solar water heaters after the late 1950s.

America's rediscovery of solar *space* heating came almost by accident. In 1932, George Keck, a Chicago architect, was asked to build the "House of Tomorrow" for the Chicago World's Fair. His design featured walls of which 90 percent was glass. Here's what Keck noticed when he dropped in on the site of the project one winter day:

> The workmen were finishing it up—this was January or February....[and] the sun was shining very brightly. There were about half a dozen workmen in their shirts, without coats on. Yet there was no artificial heat [because] they hadn't installed the heating plant. It was below zero ouside and the men were working with just their shirts on and were comfortable in the house—it was the heat of the sun!

Keck, realizing the sun's energy could replace fossil fuels for house heating, began to refine his design for solar houses. Soon solar developments emerged.

In 1940, *Business Week* called solar energy the "newest threat to domestic fuels." Unfortunately, World War II stopped residential building cold. America's building surge following the war happened too quickly for solar heating principles to be incorporated. Solar houses cost ten percent more than conventional homes. And the conservation ethic died with the end of the war. Again, solar energy was put on the back burner.

But cheap fuels are gone now, forever. A trip to the gas station, or the latest utility bill, brutally reminds us of energy's value. The energy companies are not going to provide us with all the answers. Today, nearly 90 percent of Americans believe it is their duty to help resolve our country's energy crises, according to a study conducted by the Solar Energy Research Institute.

Realtors, insulation manufacturers, builders, and everybody else can testify to the fact that America is suddenly watching its Btus. As insurance against zooming utility bills, homebuyers now choose attic and wall insulation over wall-to-wall carpeting and family rooms. Since 1973 close to 20 million homes have been retrofitted with attic insulation. A 1979 poll found that almost half of America's households planned on taking a major energy saving action during 1980 and 1981.

This drive to save energy has launched a new industry. In just three years (between 1975 and 1977) sales of solar equipment jumped from $25 million to $260 million. A 1981 government listing of solar professionals includes more than 3,500 firms and individuals.

Solar energy's greatest impact will be in new construction. A nation of solar houses isn't an idealist's pipedream. Forty million dwellings will be built between today and the year 2000. Knowledge of passive techniques has progressed so that there is no technical reason why each one can't be a solar house.

But the building stock turns over slowly. Sixty million of America's 80 million dwellings will still be around in 2000. Fortunately, estimates show that up to 60 percent of these can be adapted to use some solar energy. Thousands are already discovering the ideal solar addition—the heat-producing greenhouse.

Such other solar-based technologies as wind power also will be part of our solar future. No major technological barriers prevent widespread use of wind generators, according to Department of Energy Studies. Over 6 million rural households are in areas windy enough to make wind generation of electricity practical.

Shimmering solar cells, changing sunlight into electricity, could dot our rooftops before long. Only cost prevents solar cells from being used more now. But their price is just one-half of 1 percent of what it was just a few years ago. Prices will continue to fall. The same technology that allowed the price of pocket calculators to nose-dive benefits solar cell research.

Pathways to Growth

Until recently, solar's amazing progress coincided with massive increases in federal funding. The solar budget grew from $2 million in 1972 to more than $800 million in 1980. Many solar experts felt that too much of this money went for impractical and expensive ventures. High technology projects mimicking the space and nuclear programs ate up much of the funding. This "Big Solar" orientation led to the development of such things as the solar satellite program. If this is implemented, sixty satellites, each bigger than Manhattan Island, would orbit 24,000 miles above the earth. They would beam solar energy back to huge receiving stations. The cost—approximately $25 billion.

In Barstow, California, is a pilot-project "Power Tower." This is a boiler atop a 500 foot tower. Surrounding the tower are 2,200 giant mirrors, all beaming sunlight at it. The power tower as it is envisioned could supply electricity for an entire community.

"Those in power always want big accomplishment—scientific breakthoughs and politically viable facilities," according to M.C. Gupta, an official of the Indian Institute of Technology. Certainly our government is intrigued with the super projects. Power towers and other schemes for delivering solar energy as if it were fossil fuels are also pushed by the big energy companies. A 1976 study by Common Cause found that 73 of 139 top energy department officials were borrowed from fossil energy corporations.

Our government's priorities must be refocused. Options that are ready *now* must be further subsidized. Oil, gas, coal, and nuclear industries have received more than $200 billion in subsidies—so far. Solar has received less than 1 percent of this amount. For some reason, passive solar (the best choice for new construction) is effectively excluded from federal tax incentives.

The professional community is in a state of solar readiness. Builders, architects, and engineers have jumped on solar as a new area of necessary expertise. More than one-third of the nation's builders have used passive solar techniques. More than 75 percent of the builders who participated in the first governmental solar demonstration programn said they'd like to build more solar houses.

And building codes are changing to accomodate solar systems. This is important because there are over 10,000 municipal building codes nationwide. An unfavorable code requirement can stop a would-be solar homeowner cold. Coral Gables, Florida, at first rejected solar collectors outright. Then that decision was reversed but such rigid codes were set that costs became prohibitive.

Davis, California, took another approach. In 1972, the town council decided to reduce the energy used by Davis apartments and houses. The council hired a team of solar energy experts to examine their building code. A new code, embracing solar and energy conseration principles, was created. In all new buildings, natural or cross ventilation must be provided. Light-colored roofing materials must be used to reflect summer heat. Whenever possible, the buildings must face south. These kinds of building code modifications have allowed Davis to reduce its energy use significantly.

What do utilities think of solar? Generally, not much. Utilities view themselves as the providers of energy. They feel threatened by solar energy because solar systems give homeowners more control over their own energy supplies.

The public's enthusiasm for solar compels utilities to acknowledge that the option exists. But nothing says utilities must judge solar fairly. In a consumer pamphlet, one New Jersey utility begins by describing the sun as a "giant nuclear fusion furnace"—a description intended to reduce solar to an energy source they support. A "sociological impacts" section features a solar homeowner being verbally assaulted by his neighbor for blemishing their neighborhood with solar collectors. In a section entitled "some problems," two oldsters are shown clutching at blankets, shivering—presumably because their new-fangled solar system quit working.

However, highly centralized applications of solar energy are being eagerly pursued by utilities. The Electric Power Research Institute (funded by utilities) is working with the Department of Energy on schemes to concentrate solar cells in large fields—up to half a square mile in area. Southern California Edison is serving as project manager for "Solar One," the first U.S. power tower to provide electricity for residential use.

If too many utilities continue to discourage solar, its growth will be slow. The public must demonstrate its support for solar loudly. Government regulations can force utility cooperation. As one Houston Power and Light executive put it "when solar becomes significant, the utilities will have to become involved."

The oil companies may never come around. Unless solar energy can be turned into a centralized stock, like coal or oil, "Big Oil" has little use for it. The editor of *World Oil* had this to say regarding solar energy's future impact: "the source will have the impact, over the next quarter century, of a mosquito bite on an elephant's fanny." Exxon, in its *1980 Energy Outlook* statement, has this to say regarding solar energy's potential:

> Solar offers significant potential for the future, but its growth is presently inhibited by high cost. Technological breakthroughs may eventually bring costs down but even after these breakthroughs occur it will take decades to apply the new technology widely.

Another Exxon publication, *World Energy Outlook*, sums up Exxon's position succintly:

> Renewables such as solar may contribute importantly in the next century but they are not expected to be a major fuel source before then.

Shell, in the *National Energy Outlook (1980-1990)*, takes an almost sarcastic tone:

> Solar power boasts great potential but technical and economic hurdles will deter major contributions from solar in this decade.

Despite the poor rating they give solar, several of the world's largest oil companies, including Shell, Exxon, Arco, and Mobil have acquired all or part interest in solar cell companies. According to Richard Munson of the Solar Lobby, Exxon and Arco will soon control more than half the solar cell industry. Will the oil companies go slow on solar to maximize their profits from non-renewable resources?

Solar is best implemented in a decentralized manner. It makes little sense to gather up a resource that falls in household sized amounts all over the place, concentrate it in a single place—and then distribute it again to homeowners in household sized amounts. Decentralization means that millions of decision makers must decide to "go solar." But a recent survey found that 73 percent of homeowners don't know enough about solar energy to make a decision about using it in their homes.

There is no doubt that solar can contribute significantly to our energy supply within a generation. Just how large a contribution solar will make is hard to predict. We already have the technology necessary supply 20 percent of our national energy needs by solar, although so far we are ignoring most of that capacity. By the year 2000, experts predict that solar will supply from 7 to 40 percent of our energy needs. The answer to our energy crisis is right above us. The sun has provided energy for the earth billions of years and will continue to provide it for billions more. It will shine just as strongly tomorrow if we tap its energy today.

The World Turns to Solar Energy

In the bitter cold of Saskatchewan, Canada, a demonstration house has been built without a furnace. All of its heating needs are provided by sunlight. In Japan more than 2 million solar hot water heaters are now in use. And in Israel more than a third of all homeowners heat their water with the sun. Super-insulated solar heated houses are being created to weather the long winters of Sweden and Denmark.

Today, solar cells are the cheapest method of producing electricity for the millions of small villages that are hundreds of miles from the nearest wall socket. Passive solar dwellings can be built almost anywhere. China, which may have more passive solar buildings than any other country, makes extensive use of solar greenhouses.

More than sixty nations have formal solar research programs. Many countries now consider sunlight as the most sensible energy source for heating.

21. Energy Conservation

During the last several years any American who has had to pay a utility bill has probably had to learn something about energy conservation as well. There are two very effective ways to conserve energy in the home. The first is simply to use much less of it; set the thermostat back, especially at night and when there's no one at home. Don't leave the back door open; turn the lights off.

The second way to conserve energy is to do work on the house that will reduce the amount of heat that is lost in the winter and gained in the summer. The first step is to apply weatherstripping around doors and windows. The next is to caulk around the frames of those same windows and doors. These relatively inexpensive measures are very important, because a great deal of heat is lost by leakage through these seemingly individual cracks in the building shell.

One of the next stages is to add insulation to the attic, followed by the addition of storm windows (glass is expensive but temporary plastic storms can be used). The water heater should be insulated; the furnace should be serviced; and the possiblity of having loose insulation blown into uninsulated walls should not be ruled out.

Let's examine each of these conservation tactics, taking first things first.

1. Thermostat Set-Back: Unless your health or age demand a very warm environment, setting the thermostat back both day and night will reduce your fuel use by an amount in the neigh-

	1st Year Cost	Yearly Savings
■ **Turn Down Thermostat in Winter**	**$0**	**$20-65**
■ **Turn Up Thermostat in Summer**	**$0**	**$5-15**
■ **Put on Plastic Storms**	**$5-7***	**$15-55**
■ **Service Oil Furnace**	**$25**	**$15-40**
■ **Caulk and Weatherstrip**	**$75-105***	**$50-125**
■ **Insulate Attic**	**$160-290***	**$60-170**
TOTAL ESTIMATED COSTS	**$265-427**	**$170-470**

*These are do-it-yourself costs. If you have a contractor do it, these items could cost about twice as much.

Table 1. Initial cost, and annual saving, for simple energy conservation measures

borhood of 10 percent. In fact, the attitude of all family members toward energy can save a great deal more. If no one is going to be home for several hours, remember to set the thermostat back. Don't leave windows or doors open. *Be energy conscious*.

2. Weatherstripping: Heated air leaves a house rapidly through the cracks around windows and doors. Weatherstripping materials fastened to window and door jambs provide a tight seal. Air flow is effectively eliminated. The best and most expensive weatherstripping is called *spring metal*. Other choices are *felt, adhesive-backed foam*, and *tubular gaskets*. For the bottoms of doors there are *door sweeps* and *door shoes* to eliminate air flows.

3. Caulking: This requires a caulking gun and cartridges of caulk compound. The caulk is applied outside around the edges of door and window frames to seal the cracks through which heated air will escape. This is best done in the spring or fall to avoid temperature extremes. When purchasing caulk compound, avoid the inexpensive products that have little durability. Both silicon and butyl rubber caulks have shown good durability.

4. Insulated Water Heater: Heating water accounts for about 15 to 20 percent or more of the energy used in a house. The hot water tank should be insulated with fiberglass batt insulation. There are special kits available for this purpose. Most water heaters have thermostat controls, and these can be set back to 100°F or 120° from the 140°F-plus where they are usually set. (But first make sure that your dishwasher does not require the higher temperatures.) The hot water burner should be serviced once a year to maintain efficiency.

5. Service the Furnace: If you use oil or gas heat, the furnace should be serviced and cleaned by a heating technician before the start of each heating season. This will improve the efficiency and decrease the amount of fuel burned. Replacing older burners with newer, more efficient types is even more of an improvement.

6. Attic Insulation: The attic is a major point of heat loss. It should be insulated with *at least* 6 inches of fiberglass batt, preferably more. Also be sure to insulate the attic door.

7. Storm Windows: Heat is lost rapidly by conduction through windows. Storm windows, in conjunction with caulking and weatherstripping, make window units far less detrimental to comfort and budget. Temporary plastic storms are popular substitutes for real storm windows, which are fairly expensive.

8. Wall Insulation: If the walls of a house are not insulated, then it is worth considering having loose cellulose or fiberglass blown into the wall cavity. The other alternatives are either to have foam insulation blown in (expensive and a potential health and/or fire hazard), or to tear the interior finish wall out in order to install fiberglass batt insulation and a vapor barrier, or to remove the siding of the house so that rigid board insulation can be installed. In all cases, pay serious attention to the vapor barrier that prevents heat-driven moisture from inside from getting into the insulation and lessening its value. Even two coats of gloss or semi-gloss paint put on interior walls before the finish coat adds a lot of protection.

22. The Underground House

Figure 1. Earth-sheltered house built into a south-facing slope

Going underground is yet another response to the inflated price of energy. Below the frost line, the constant 45°F to 50°F ground temperature provides a year-round buffer against temperature extremes. A house is considered to be underground or earth-sheltered if it is built entirely or partially beneath the surface, or if it is built above ground with earth banked up against one or more of its sides. This latter procedure is called earth-berming.

Building underground also provides protection against wind, and reduces the amount of heat lost by air leaking through walls and building joints.

An underground house does need to overcome the feeling of a dark and damp living space. Naturally, the best way to accomplish this is to keep the south-facing side of the house open to the sun. This will allow for passive solar heating and an abundance of light.

These buildings also have other advantages. Because they are built with heavy, masonry construction, they tend to have long life-expectancies, low maintenance, fire resistance, and increased comfort because temperature swings are minimal and there are few drafts. Their construction is, however, more demanding than for above-ground structures; the services of an architect or engineer are usually required during the planning stages.

There are several other important things to consider when planning an earth-sheltered building:

Site

The building site is central to the planning of an earth sheltered home. Soil and ground water conditions will determine structural and waterproofing requirements. For example, some soils are more susceptible than others to expansion when wet or frozen; these will place more demands on the strength of the building. An engineer or soil-testing firm may be needed at this critical stage in the planning and design. The topography of the site—"the lay of the land"—will affect windflow and drainage patterns, and will determine how easily the building can be surrounded with earth. A modest slope requires more excavation than a steep one, and a flat site is the most demanding, needing extensive excavation. Buildings on flat ground are more easily "bermed" on one or more sides. Berming is the practice of banking earth up against the walls of the building. A south-

Figure 2. Earth-sheltered house with a central court

facing slope is ideal for an earth-sheltered building because the southern wall can be left exposed to the sun for direct heating through windows while the rest of the house is set back into the slopes (see *Figure 1*). Every site differs, of course, but most have features that can be put to use.

Design

While it makes the best use of the building site, an earth sheltered house still can be varied in design to suit the tastes of the occupant. When a house is built almost entirely underground, the first consideration is to provide natural light and solar heat to the living spaces. The floor plan is arranged so that the main spaces share light and heat from the southern exposure (once again, an exposed, glazed south-facing wall is an excellent approach). This can be modified by building a greenhouse along the south wall.

Another approach is the central court (see *Figure 2*). This allows rooms to surround an outdoor space on three or four sides. The strategic use of clerestories and skylights will also allow more latitude in the arrangement of interior spaces. The design should counteract the negative efect of being underground. A properly executed design will leave the occupant with the feeling that there is very little difference from living above the ground.

It is important to provide fire escape routes from bedrooms; building codes require this.

Construction Materials

The materials used to build earth-sheltered houses must be able to withstand the stress imposed by the surrounding earth. When soil is wet or frozen, for instance, it exerts greater pressure on the walls and floors of the building. Pressure also increases with depth, so materials must be selected and used accordingly. The common building materials—concrete and reinforced masonry, wood, and to some extent, steel—are all eligible.

Concrete is usually the first choice for construction of earth sheltered buildings. It has the added advantages of durability and fire resistance. Unreinforced concrete, poured at the site, usually is used for footings and floor slabs, and can be used for walls at shallow depth. Reinforced concrete, on the other hand, has the ability to resist loads at any reasonable depth and can be used for floors, walls, and roofs.

Concrete will absorb and store solar heat as part of a passive system, and its heat-absorbing qualities help to prevent large temperature swings.

Pre-cast concrete has the advantages of poured concrete and more. It meets all structural requirements, and construction proceeds more quickly with pre-cast units. Special care must be taken, however, in making the joints between sections water tight.

Concrete block, surface bonded with fiberglass, can be used for walls up to two stories but needs reinforcement at depths greater than 6 to 8 feet. Cracks in mortar joints must be sealed carefully. The porous quality of block demands extra care

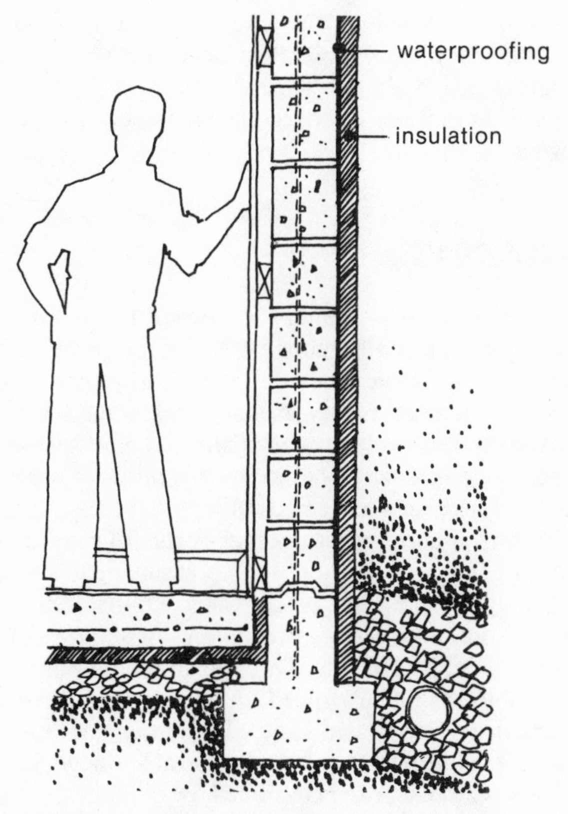

Figure 3. Construction details for earth-sheltered houses

during the waterproofing process.

Wood has been used for walls and roofs of earth-sheltered buildings. It is less expensive than concrete but is not as strong. It needs to be pressure-treated with preservatives that will enable it to endure moist conditions. Conventional framing techniques, specially adapted, can be used, and post and beam construction has also proven successful.

Steel can be used for beams and column supports, but it has been used in other less conventional ways as well. For example, large steel culvert sections have been used to form the shell for dwellings, finished on the interior and waterproofed on the exterior. They have more than adequate strength: they were designed originally for underground use and are also quite impervious to water seepage. They must, however, be treated to prevent corrosion.

Waterproofing

Waterproofing is extremely important to an earth-sheltered house, and it must be done right the first time so that major excavation is not necesary to locate and repair leaks. Underground structures must withstand prolonged periods of water pressure. The common waterproofing techniques used for basements, such as coating the exterior walls with asphalt, are not suitable for earth-sheltered dwellings.

The first step in waterproofing occurs during the site selection. The best way to avoid water pressure against underground walls is to choose a site where water will naturally drain away from the building. Survey the site for low spots and areas where water will collect. If possible, build above the water table. If the water table cannot be avoided, drainage systems will have to be designed to draw water away. Building at the top of a hill is usually a good choice because of natural drainage, but a percolation test is advisable to determine how quickly the soil transfers moisture. Earth-bermed houses have minimal water problems because they are built on or just below ground level, and moisture in the berms drains off by gravity.

After drainage systems have been designed, the actual water proofing materials for the building can be selected. The limited amount of available data available does not indicate that any one approach is "best," but the selection should meet three important criteria: (1) it should have a long life expectancy underground; (2) it should have resealing capability at underground temperatures; and (3) it should have good crack-bridging capability. Backfilling and construction imperfections can cause undetectable punctures in the waterproofing material so its ability to reseal is important. Masonry walls always crack during curing and settling, and the waterproofing must be able to bridge these cracks adequately to prevent leaks. The life of different materials can only be predicted, and there are few, if any, test cases old enough to demonstrate long-term durability.

The products most widely used for waterproofing systems at present are: asphalt and pitch-impregnated membranes; rubberized asphalt; butyl rubber and EPDM (ethylene propylene diene monomer) membranes; liquid polyurethane; and bentonite.

Built-up asphalt or pitch membranes have been used sucessfully in underground structures for both roofs and walls. The felt products normally used in built-up roofs are not recommended because they tend to rot under prolonged exposure to moisture. Built-up membranes also lack good resealing expansion and crack bridging qualitities. Fiberglass fabric should be used instead.

Bituthene is rubberized asphalt coated with polyurethane. It can be applied directly to walls and roofs and has a long life expectancy.

Butyl rubber and EPDM membranes are durable rubber sheets that are glued to walls or roofs of the building and have been successful where applied properly. The point of failure is likely to be at seams that are not sealed properly.

Liquid polyurethanes are often used at points where it is difficult to apply a membrane. They have no seams, which is an advantage; however, they do not reseal. Polyurethanes are sometimes used as coatings over insulation on underground structures.

Bentonite is a natural clay that can be formed into panels or applied as a liquid spray. The panels are simply nailed to roofs or walls. They expand when wet and seal out moisture. The spray is mixed with a mastic and applied ⅜-inch thick to roofs or walls.

Insulation

Insulation in underground buildings is as necessary as it is in conventional buildings, but there are some special considerations. If the building is of masonry construction, insulation usually is placed on its exterior (see *Figure 3*). This allows the solar energy that is collected and absorbed by the concrete or block to be retained within the building. In most cases the waterproofing is applied to the building first—before insulation. Because the insulation will be exposed to the earth, it should be a closed-cell product (usually a rigid board) like extruded polyurethane that can also withstand the pressure of back-filling without being compressed.

Building codes and regulations vary quite a bit from state to state and from one muncipality to another. Before beginning any construction project it is usually necessary to obtain the appropriate building permits. Because earth-sheltered houses have not been considered in most building codes, they may present some special problems to local code officials.

Some typical obstacles are found in the Uniform Building Code, which is published by the International Conference of Building Officials as a model document and has been adopted as a basis for code writing in many local jurisdictions throughout the United States. For example, the UBC requires that habitable rooms have glazed areas greater than 1⁄10 of their floor area. This is impossible for a room surrounded on all sides either by earth or by other rooms. Another requirement is that bedroooms have a window or door leading to the outside both for quick exit and for access to fresh air to prevent smoke inhalation in the case of fire. There may be alternate ways to provide safety for occupants, such as smoke detectors or well-planned escape routes, but the burden of the proof is on the owner or builder.

Because of such code barriers, it is important that a builder of earth-sheltered houses be familiar with the local requirements and he or she work with local officials in the planning stages to prevent problems or to provide solutions before work begins.

23. The Solar Cell

Of all the technologies invented by humans, one is so remarkable, yet so simple, that it is still difficult to appreciate fully. That technology is the solar cell, also called the *photovoltaic* cell: *photo* as in *photon*, the elementary particle of sunlight, and *voltaic* as in voltage or the flow of electrons we call electricity.

Solar cells convert sunlight to electricity, directly. Nothing is burned or boiled. There are no waste products or pollutants.

These cells can be made from a number of substances, but silicon—the Earth's second most common element; the most familiar example is sand—is used most frequently. Two thin slices of a highly purified, carefully "grown" silicon crystal are joined together to produce a single cell. One slice has a positive charge and the other a negative charge. When sunlight strikes a cell, the photons jar electrons loose from the negative layer. This starts the flow of electrons.

Solar cells were first developed by Bell Labs in 1954 and used most prominently by NASA in the space program. Their original cost was so great that the idea of using them to produce electricity for the everyday world was considered rather far-fetched. But the costs have dropped dramatically, and it is now reasonable to believe that solar cells can become an economical form of power production in the not-too-far distant future.

Research and development of cheap cells with higher efficiencies has been stepped up over the last decade. Mass production techniques are being developed. Markets are being cultivated. But there are problems.

A great deal of power and money has been committed to the production of electricity by other means. For years nuclear power has ben touted as the future source of electricity. But nuclear is now facing the harsh music of public mistrust, growing economic unviability, and the ever-present fear of catastrophe. While the investors in nuclear continue their attempts to salvage it from public rejection (witness the sugary advertisments placed on the Op-Ed page of the *New York Times*), solar cell development is left lingering in the background.

But some experts predict that by the late 1980s solar cells will be cheap enough to be placed on the roofs of houses. Some experimental installations have already been made. The electricity produced can either be stored in batteries or fed into the utility electical grid so that the homeowner gains credit for his site-produced power. Thè electric meter could actually run backwards.

The potential demonstrated by solar cells for bringing electricity to the villages of the underdeveloped countries may promise one of the best solutions to the energy problems of the Third World. Schuchuli, a village on the huge Papago Indian reservation in southeastern Arizona, was one of the first communities in the world to rely entirely on solar energy for its power needs. The village, with a population of 95, is 17 miles from the nearest available electric utility. Previous to the installation of the photovoltaic system on December 16, 1978, Schuchuli's lighting was provided by kerosene lamps and by candles. Perhaps this step will serve as a model for the nonindustrialized world.

The largest solar cell generating unit in the world is found at Mt. Laguna, California. The 60-kilowatt power system uses close to 97,000 silicon solar cells to provide power for the Air Force station located there.

There are now at least twenty companies in the United States capable of manufacturing solar cells, including subsidiaries of four of the world's largest oil companies, Mobil, Exxon, Gulf, and Arco. As the fortunes of nuclear and oil-generated electricity fall, we can all watch with interest as those of solar cell power rise.

Some of the most active manufacturers of solar cells in the United States are:

ARCO Solar, Inc
20554 Plummer St.
Chatsworth, CA 91311

Mobil Tyco Solar Energy Corp
16 Hickory Dr.
Waltham, MA 12154

Solarex Corporation
1335 Piccard Dr.
Rockville, MD 20850

Solar Power Corp
20 Cabot Rd.
Woburn, MA 01801

Texas Instruments, Inc
Post Office Box 5012
Dallas, TX 75222

24. Solar Satellites

Solar collectors on earth must suffer the indignity of idleness between sunset and sunrise. They will do the job when the sun is out and, if they are properly coordinated with a storage component, their energy can be stretched to cover the sunless hours as well.

But there are dreamers among us who are having rather big dreams. The dreamers, like Dr. Peter Glaser of the Arthur D. Little research firm and Gerard K. O'Neil of Princeton University, envision solar power as something best developed in outer space. They would like to see huge satellites—as large as 72 square miles—in stationary orbit around the earth. These satellites would be covered with solar cells (see *Solar Cells*, page 129).

The solar cells would be exposed to the sun for almost 24 hours a day and would be continually converting sunlight into electricity. The electricity would in turn be converted into microwaves that would be transmitted to large receivers on the earth, where they would be reconverted and distributed through the existing power lines.

One 72-square-mile satellite might supply electricity to one million homes. And the dreamers see hundreds of these satellites orbiting the earth, changing the face of the sky to include even more man-made "stars."

The best place to manage the construction of solar satellites would be from the moon, where reduced gravity would make launching the materials needed to build the power stations far

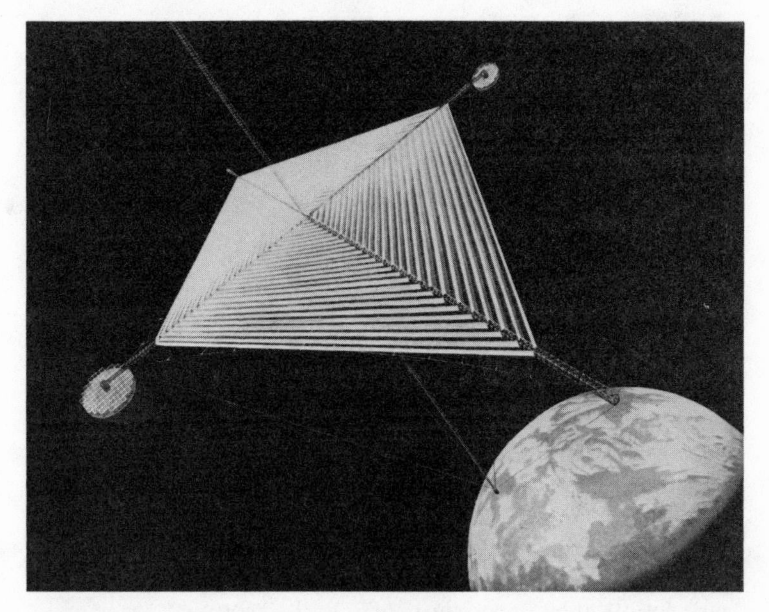

An artist's conception of a completed solar satellite in orbit

easier and less damaging to earth's atmosphere. The satellites would be assembled in space. The lunar bases would require a worker's colony built around a complex of mines, factories, and launch facilities. The moon would serve as the source of much of the needed raw materials used in satellite construction.

Opponents of the plan point out the sizable technical and economic difficulties—difficulties that include the possible environmental and human impact of the use of microwaves and the addition of energy from space to the earth's atmosphere, the enormous amount of money the plan would cost, and the possibility that the energy cost of constructing and maintaining the satellites might exceed the energy they could produce. Supporters point to the need to begin industrializing space so that greater prosperity can be brought to the earth. They assert that because there is no gravity in space it is the perfect location to manufacture highly purified substances and precision instruments for use in high technology. For them, the solar power potential is only the immediate justification for an adventure that could lead to elaborate space colonies and industries. Such a step would alleviate some of the stress that increasing earth bound industry places on the biosphere.

Solar satellite proponents also insist that there is no need for any profound technological breakthrough. They say that all this is possible through the extension of existing technology, particularly with the advent of regular trips by the Space Shuttle.

The opponents have the edge in the debate so far: the cost of solar satellites *is* stupendous. With an initial investment of $40 billion just to get started, the return on that commitment would be slow, if it ever came at all. But perhaps as time goes on the supporters of solar satellites will prepare more convincing arguments that will rally both private industry and government to their cause. Or perhaps a political leader will come along to put his or her mark on the future industrialization of space, much the way John Kennedy set our course to the moon.

Certainly the dreamers cannot be faulted for their vision, but what we have here is an undertaking analogous to what the pyramids must have been in their time or to what the voyage of Columbus must have meant in 1492, when his opponents warned him about the horrible fall he was going to take off the edge of the earth.

25. Wind Energy

"One man's meat," the saying goes, "is another man's poison." Wind energy on a small scale is kind of like that. It is best suited to rural and out-of-the-way places, but tends to get sticky in the close quarters of a village or the suburbs. It can be used best as an energy source for single buildings, or small groups of buildings—or, if the wind generators are in groups, as "farms" for particular locations.

Wind energy is almost always more expensive than standard utility-produced electricity. But its charm—its use as a symbol of independence and self-reliance—is what wind energy is all about. Like other solar technologies, but even more so, wind power is something for the adamant individual.

A wind system has four major components. These are: the wind machine itself, the tower that supports the machine, the storage batteries (or a connection to the utility power line), and the electrical equipment—such as voltage regulators, inverters, and automatic switches—necessary for the system. Together these four components involve a complex technology with an elaborate jargon to go with it.

The fundamental requirement for a wind machine is a site with good consistent wind. The machine must be *at least* 40 feet above the ground, and 15 to 20 feet higher than any obstructions. An obstruction—a building, a tall tree, a hill—can disturb the wind for several hundred yards behind the obstacle and sometimes even 50 to 100 yards in front of it. This requirement alone places limitations on the chances for having a successful wind energy system.

A wind machine can be expected to provide electricity for appliances and lighting but it should not be considered as a source for heat.

The costs for wind machines designed to provide electricity to a house or small business with similar needs might range from $5,000 to $10,000. The investment will pay for itself in ten to twenty years. The costs include the equipment, installation, maintenance, and financing.

Some of the principal manufacturers of wind energy machines, from whom you can get product descriptions, prices, and in some cases the names of qualified installers are:

Aeropower, Inc
2398 4th St.
Berkeley, CA 94710

Kedco, Inc
9016 Aviation Blvd.
Inglewood, CA 90301

North Wind Power Co
Box 315
Warren, VT 05674

Pinson Energy Corp
Box 7
Marston Mills, MA 02648

Sencebaugh Wind Electric
Box 11174
Palo Alto, CA 94306

Grumman Energy Systems
4175 Veterans Memorial Hwy.
Ronkonkoma, NY 11779

26. Further Sources of Information

This state-by-state listing of agencies and organizations can provide further regional and local information about solar energy. Many of these sources will be able to name solar professionals in your area as well as give you information about solar education and training programs. They will also have information about solar tax benefits, solar access laws, and land use. Many of the sources operate toll-free 800 energy hot-lines.

Alaska

Information:
Department of Commerce & Economic Development
Division of Energy and Power Development
338 Denali St.
Anchorage, AK 99501
(907) 276-0508
Conducts policy planning, alternative energy studies, wind demonstration projects, surveys and workshops; maintains solar advisory group; distributes consumer information.
Other organizations:
Alaska Center for the Environment, 1069 W. 6th Ave., Anchorage, AK 99501.

Alaska Energy Extension Service, 338 Denali St., Anchorage, AK 99501.

Alabama

Information:
Alabama Solar Energy Center
c/o Kenneth E. Johnson Center
University of Alabama/Huntsville
Huntsville, AL 35807
(800) 572-7226
Tests collectors and systems; conducts workshops and exhibits; maintains information service and library; answers solar questions; distributes own test reports and studies, state solar directory, and other publications.
Other Organizations:
Alabama Department of Energy, Washington Court, 25 Washington Ave., Montgomery, AL 36130.
Alabama Energy Extension Service, University of Alabama, Natural Resource Center, PO Box 6282, University, AL 35486.
Alabama Solar Coalition, 1750 Flower Wood Dr., Birmington, AL 35216.
Alabama Solar Energy Association, c/o University of Alabama Environmental & Energy Center, PO Box 1247, Huntsville, AL 35807.
Safe Energy Alliance, c/o Bonnie Britton, Rt 2, PO Box 347, Auburn, AL 36830.

Arizona

Information
Arizona Solar Energy Research Commission
Rm. 502-Capitol Tower
1700 W. Washington
Phoenix, AZ 85007
(800) 352-5499
(602) 255-3303

Conducts policy planning, feasibility studies, state demonstration projects, conferences and workshops (for general public and solar design professionals). Distributes state solar directory, consumer guide, and federal solar publications.
Other organizations:
Arizona Energy Extension Service, 1700 W. Washington Street, Phoenix, AZ 85007
Arizona Solar Energy Association, P.O. Box 1443, Flagstaff, AZ 86002

Arkansas

Information
Arkansas Energy Office
3000 Kavanaugh Blvd.
Little Rock, AR 72205
(501) 371-1370 Solar information
(800) 482-1122 State Energy Hotline
Answers solar and general energy-related questions; distributes fact sheets and bibliographies. Conducts workshops for general public and civic organizations.
Other organizations:
Arkansas Energy Extension Service, 3000 Kavanaugh Blvd., Little Rock, AR 72205
Arkansas Solar Coalition, 1145 W. Hearn, Blytheville, AR 72201
Mo-Ark Solar Energy Association, Box 1643, Jefferson City, MO 65102

California

Information:
California Energy Commission
1111 Howe Ave., Mail Station 70
Sacramento, CA 95825
(800) 852-7516 State Solar Hotline

Answers solar and other alternative energy questions, state solar tax credit inquiries, consumer complaints. Tests and certifies collectors; conducts policy planning; develops code manuals for buildings and planning officials. Distributes list of California solar professionals, information on state solar tax credit, and solar publications.

Other organizations:

California Department of Consumer Affairs, Solar/Insulation Unit, 1020 N St., Room A547-F, Sacramento, CA 95814
California Energy Extension Service, 1211 16th Street, Sacramento, CA 95814
California Office of Appropriate Technology, 1530 10th Street, Sacramento, CA 95814
California Solar Energy Association, 202 C Street, San Diego, CA 92101
California Solar Energy Industries Association, c/o Agnes James, 926 J Street, Building 1021, Sacramento, CA 95814
Foothill Solar Exchange, 105 Rockwood Drive, Grass Valley, CA 95945
Northern California Solar Energy Association, P.O. Box 1056, Mountain View, CA 94042
Pacific Alliance, P.O. Box 1738, San Luis Obispo, CA 93406
Santa Cruz Alternate Energy Co-op, P.O. Box 66959, Scotts Valley, CA 95066
SolarCal Council, State Capitol, Sacramento, CA 95814
SUNRAE, 5679 Hollister Avenue, Rm. 5B, Goleta, CA 93017

Colorado

Information

Office of Energy Conservation
1600 Downing Street, 2nd Floor
Denver, CO 80218
(303) 839-2507

Conducts policy planning: develops educational programs (vocational and other levels); conducts workshops. Distributes solar information and publications of state-wide interest. Offers information on regional energy extension centers.

Other organizations:
Alternate Energy Society of Pueblo, 4015 Hillside Drive, Pueblo, CO 81008
Boulder Solar Energy Society, P.O. Box 3431, Boulder, CO 80307
Colorado Solar Energy Association, P.O. Box 5272, Terminal Annex, Denver, CO 80217
Colorado Energy Extension Service, 1600 Downing Street, Denver, CO 80218
Durango Energy Conservation Center, P.O. Box 1948, Durango, CO 81301
Grand Junction Solar Energy Association, c/o Public Energy Information Office, 250 N. 5th Street, Grand Junction, CO 81501
Huajatolla Primal Energy Association, P.O. Box 467, La Veta, CO 81055
Roaring Fork Resource Center, P.O. Box 9950, Aspen, CO 81611
New Age Energy Council, 750 Whedbee Street, Fort Collins, CO 80524
People's Alternative Energy Service, Rte.1, Box 3A, San Luis, CO 81152
San Luis Valley Solar Energy Association, P.O. Box 1284, Alamosa, CO 81101
Solar Energy Association of Northeastern Colorado, P.O. Box 307, Eaton, CO 80615
Southeast Colorado Alternate Energy Association, 318 Santa Fe, La Junta, CO 81050
Western Slope Energy Research Center, P.O. Box 746, Hotchkiss, CO 81416
Western Solar Manufacturers Association, 4785 Elati Street, Suite 18, Denver, CO 80216

Connecticut

Information:
Office of Policy and Management Energy Division
80 Washington Street
Hartford, CT 06115
(203) 566-2800 Solar information
(800) 842-1648 State Energy Extension Service
Conducts policy planning, feasibility studies, offers information on state solar tax incentives, courses and consumer protection. Offers Connecticut Energy Outlook, an assessment of the state's energy supply and demand (with recommendations for legislative action), in summary form.
Other organizations:
Connecticut Energy Extension Service, 80 Washington Street, Hartford, CT 06115
Solar Energy Association of Connecticut, P.O. Box 541, Hartford, CT 06101

District of Columbia

Information:
Office of Planning and Development
Energy Unit
1329 E. Street, S.W.
Rm. 759
Washington, D.C. 20004
(202) 727-1800
Other organizations:
Anacostia Energy Alliance, 2027 Martin Luther King, Jr. Ave., S.E., Washington, DC 20020
District of Columbia Solar Coalition, 236 Massachusetts Ave., N. W. , Suite 610, Washington, DC 20002
Institute for Local Self Reliance, 1717 18th Street, N.W., Washington, DC 20009

Delaware

Information
Governor's Energy Office
56 The Green
Dover, DE 19901
(302) 736-5644
(800) 282-8616
Conducts policy planning, feasiblity studies, certifies systems for state solar tax credit; offers training programs for installers and building inspectors, distributes material for public schools. Administers state-funded renewable resource grants program.
Other organizations:
Delaware Energy Extension Service, 56 The Green, Dover, DE 19901
Delaware Energy Office, DT & CC-South, Rm. 301, Georgetown, DE 19947
Delaware Solar Energy Association, RD #3, P.O. Box 289K, Oxford, PA 19363
Delawareans for Energy Conservation, 6 Park Lane, Dover, DE 19901

Florida

Information
Florida Solar Energy Center
300 State Road 401
Cape Canaveral, Fl 32920
(305) 783-0300
Conducts technical and market research, tests collectors, conducts training programs for installers. Distributes consumer and technical publications, answers solar questions.
Other organizations:
Florida Solar Coalition, 935 Orange Avenue, Winter Park, FL 32789
Florida Solar Industry Association, Executive Association Services, 1300 Executive Center Drive, Tallahassee, Fl 32301

Florida Solar Users Network, 1086 Colorado Drive, Rockledge, FL 32955
State Energy Office, 301 Bryant Building, Tallahassee, FL 32301

Georgia

Information
Georgia Office of Energy Resources
Rm. 615
270 Washington Street, S.W.
Atlanta, GA 30334
(404) 656-5176
Conducts policy planning, answers solar and energy related questions, maintains solar energy library, circulates a slide show of Solar in Georgia, and distributes Residential Solar Heating Guide, State Solar Directory, lists of solar installations in Georgia, and various federal publications.
Other organizations:
Georgia Energy Extension Service, 270 Washington Street, S.W., Atlanta, GA 30334
Georgia Solar Coalition, 1103 Euclid Avenue, N.E., Atlanta, GA 30307
Georgia Solar Energy Association, P.O. Box 32748, Atlanta, GA 30332
Solar Energy Industries Association of Georgia, c/o Grumman Corporation, 1420 First National Bank Tower, Atlanta, GA 30303

Hawaii

Information
State Energy Office
Department of Planning & Economic Development
P.O. Box 2359
Honolulu, HI 96804
(808)538-4150

(808) 548-4080 State Energy Hotline (on Oahu)
Dial 0, ask for 8016 (toll free within Hawaii—off Oahu)
Conducts policy planing workshops, conferences and seminars; manages state conservation programs and promotes the use of solar technologies and conservation in state buildings. Develops and distributes publications on solar energy and conservation.
Other organizations:
Center for Science Policy & Technology Assessment, Department of Planning & Economic Development, P.O. Box 2359, Honolulu, HI 96804
Hawaii Energy Extension Service, 1164 Bishop Street, Suite 1515, Honolulu, HI 96804
Hawaii Natural Energy Institute, Holmes Hall, Rm. 240, 2540 Dole Steet, Honolulu, HI 96822
Hawaii Solar Energy Association, P.O. Box 23350, Honolulu, HI 96822

Idaho

Information
Idaho Office of Energy
State House
Boise, ID 83720
(208) 334-3800
Answers solar and other energy-related questons.
Other organizations:
Idaho Energy Extension Service, State House, Boise, ID 83720
Solar Energy Association of Idaho, P.O. Box 2761, Boise, ID 83707

Illinois

Information
Institute of Natural Resources
Solar Section
325 W. Adams
Springfield, IL 62706
(217) 785-2800
Offers technical assistance for solar design; has computerized F-chart available for analysis of solar hot water. Reviews solar loan applications, conducts policy planning, feasibility studies; develops vocational education materials, and sponsors design and installer workshops. Answers solar and other energy-related questions. Ask for a copy of the center's publication list.
Other organizations:
Central Illinois Solar Energy Society, P.O. Box 170, Rochester, IL 62563
Illinois Energy Extension Service, 325 W. Adams Street, Springfield, IL 62706
Illinois Energy Resources Commission, 612 S. Second Street, Springfield, IL 62706
Illinois Solar Contractors Association, P.O. Box 175, Mossvile, IL 61552
Illinois Solar Energy Association, P.O. Box 1592, Aurora, IL 60507
Illinois Solar Energy Industries Association, P.O. Box 175, Mossville, IL 61552
Northern Illinois Solar Energy Association, P.O. Box 352, Argonne, IL 60439
Shawnee Solar Project, 211 ½ W. Main Street, Carbondale, IL 62901
Solar Resource Advisory Panel, Institute of Natural Resources, Rm. 30, 325 W. Adams, Springfield, IL 62706
South Central Illinois Solar Energy Association, 637 Eccles, Hillsboro, IL 62049

Indiana

Information
Indiana State Solar Office
Indiana Department of Commerce Energy Group
4490 N. Meridian Street
Indianapolis, IN 46204
(317) 232-8940
Conducts policy research, answers general energy-related questions. State Solar Office chartered to provide information, education and training in solar and related alternative energies.
Other organizations:
Alternative Technologies Association, P.O. Box 27246, Indianapolis, IN 46227
Hoosier Solar Energy Association, P.O. Box 44448, Indianapolis, IN 46204
Indiana Energy Extension Service, 7th Floor, 440 N. Meridian Street, Indianapolis, IN 46204
Indiana Solar Energy Coalition, Rt. 1 8244 Colt Drive, Plainfield, IN 46168
Ohio Valley Solar Society, P.O. Box 2375, Evansville, IN 47714
Richmond Solar Interest Group, c/o Richmond Energy Office, 50 N. Fifth Street, Richmond, IN 47374
Solar Collective of Northeast Indiana, P.O. Box 12891, Fort Wayne, IN 46866
Wabash Valley Solar Society, 1900 Park Ave, N. Terre Haute, IN 47805

Iowa

Information
Solar Office
Energy Policy Council
Capitol Complex
Des Moines, IA 50319
(515) 281-4420

Conducts policy planning, coordinates workshops, administers state solar research and development projects, develops energy education materials. Distributes state solar directory, annual summary of state and federal solar programs, survey of solar legislation, and selected federal publications.

Other organizations:

Citizens United for Responsible Energy, 3500 Kingman Blvd., Des Moines, IA 50311

Community Action Research Group, Box 1232, Ames, IA 50010

Iowa Center for Self Reliance, 3500 Kingman Blvd, Des Moines, IA 50311

Iowa Energy Extension Service, 110 Marston Hall, Iowa State University, Ames, IA 50011

Iowa Solar Energy Association, P.O. Box 68, Iowa City, IA 52244

Iowa Solar Energy Association, 1433 Woldwood Drive, N.E., Cedar Rapids, IA 52402

Solar Resource Advisory Panel, c/o Dr. Laurent Hodges, 2115 Coneflower Court, Ames, IA 50010

Kansas

Information

Kansas Energy Office
241 W. Sixth Street
Topeka, KS 66603
(800) 432-3537

Answers general energy-related questions (including solar). Assists residents in filing applications for federal, energy related grant programs. Distributes state solar directory and federal solar publications.

Other organizations:

Kansas Energy Extension Service, 214 W. Sixth Street, Topeka, KS 66603

Kansas Solar Energy Association, P.O. Box 8516, Wichita, KS 67208

Mid-American Coalition for Energy Alternatives, 5130 Mission Road, Shawnee Missions, KS 66205

Kentucky

Information

Bureau of Energy Management
Kentucky Department of Energy
P.O. Box 11888
Iron Works Pike
Lexington, KY 40578
(800) 372-2978 State Energy Hotline
Conducts policy planning, seminars and workshops. Answers solar and other energy-related questions. Maintains as part of the departmental library a section containing solar information, both general and technical, which is open to the public. Published do-it-yourself solar information and distributes federal solar publications.

Other organizations:

Appalachian Community Services Network, Resource Coordination Center, University of Kentucky, Lexington, KY 40506
Appalachia Science in the Public Interest, P.O. Box 298, Livingston, KY 40445
Central Kentucky Community Action Council, 225 N. Woodlawn, Lebanon, KY 40033
Environmental Alternatives, 818 E. Chestnut Street, Louisville, KY 40204
Kentuckiana Solar Society, P.O. Box 974, Louisville, KY 40201
Kentucky Energy Extension Service P.O. Box 11888, Lexington, KY 40578
Kentucky Solar Coalition, 728 Garrard Street, Covington, KY 41011
Mountain Community Energy Alternatives, Rt. 1, Waco, KY 40385
Safe Alternatives for Energy, P.O. Box 111, Lexington, KY 40511

Louisiana

Information
Research and Development Division
Department of Natural Resources
P.O. Box 44156
Baton Rouge, LA 70804
(504) 342-4594
Conducts state-wide renewable energy policy planning, administers state solar research projects. Disseminates solar information.
Other organizations:
Louisiana Energy Extension Service, Louisiana State University Cooperative Extension Service, Knapp Hall , Baton Rouge, LA 70803

Maine

Information
Maine Office of Energy Resources
State House Station 53
Augusta, ME 04333
(207) 289-2195
Conducts policy planning; administers state solar hot water grant programs. Distributes federal solar publications.
Other organizations:
Maine Audubon Society, Gisland Farm, 118 Old Route One, Falmouth, ME 04105
Maine Energy Extension Service, 295 Water Street, Augusta, ME 04333
Maine Solar Energy Association, 24 Gof Street, Auburn, ME 04210

Maryland

Information
Maryland Energy Policy Office
Suite 903
301 W. Preston Street
Baltimore, MD 21201
(301) 383-6810 solar information
(800) 494-5903 Home Energy Savers Program
Offers information on state solar legislation. Distributes lists of solar professionals.
Other organizations:
Maryland Energy Extension Service, Suite 903, 301 W. Preston Street, Baltimore, MD 21901
Solar Action of Maryland, 333 E. 25th Street, Baltimore, MD 21218

Massachusetts

Information
Massachusetts Executive Office of Energy Resources
Department of Renewable Resources, Solar Division
73 Tremont Street
Boston, MA 02108
(617) 727-7297
Conducts policy planning, feasiblity studies and workshops. Develops guidelines for building codes and property tax assessments; handles consumer problems (in cooperation with state attorney general's office) , answers technical and non-technical questions, distributes state and federal solar publications.
Other organizations:
Berkshire Solar Energy Association, c/o Mountain Craft Builders, 115 South Street, Pittsfield, MA 01201
Mass Bay Solar Energy Association, 48 Golden Ball Road, Weston, MA 02193
Massachusetts Energy Extension Service, 73 Tremont Street, Boston, MA 02108

Massachusets Public Interest Research Group, 120 Boylston Street, Boston, MA 01226
Western Massachusetts Solar Energy Association, c/o Cooperative Extension Service, University of Massachusetts, Energy Education Center/Tillson Farm, Amherst, MA 01002
Urban Solar Energy Association, 21 Burnside Avenue, Somerville, MA 02144

Michigan

Information
Solar Office
Business and Renewable Resources Section
Energy Administration
Department of Commerce
P.O. Box 30228
Lansing, MI 48909
(800) 292-4704 State Energy Information
Conducts policy planning, conferences and workshops, and information collection and dissemination.
Other organizations:
Michiana Solar Roundtable, 100 Chapel Road, Niles, MI 49120
Michigan Energy and Resources Research Association, 1200 6th Street, Detroit, MI 48106
Michigan Energy Extension Service, Energy Administration, Department of Commerce, P.O. Box 30228, Lansing, MI 48909
Michigan Solar Energy Association, 201 E. Liberty Street, Ann Arbor, MI 48104
Solar Resource Advisory Panel, P.O. Box 7323, Ann Arbor, MI 48107

Minnesota

Information
Minnesota Solar Office
980 American Center Bldg.
150 E. Kellogg Blvd.
St. Paul, MN 55101
(612) 296-5175 Solar Information
(800) 652-9747 State Energy Hotline/Solar Information
Alternative Sources of Energy, 107 South Central, Milaca, MN 56353
Center for Local Self-Reliance, 3302 Chicago Avenue, Minneapolis, MN 55407
Minnesota Energy Extension Service, 980 American Center Bldg., 150 E. Kellogg Blvd., St. Paul, MN 55101
Solar Resources Advisory Panel, P.O. Box 9815, Minneapolis, MN 55440
Minnesota Solar Energy Association Inc., 2829 University Ave., S.E., Minneapolis, MN 55414

Mississippi

Information
Mississippi Department of Energy and Transporation
510 George Street
Watkins Bldg.
Jackson, MS 39202
Monitors solar lumber-drying project, studies "solar barriers" and co-sponsors "solar barriers" workshops open to the public. Distributes federal solar publications and other energy information. Works with Missouri Solar Council to hold workshops, and promotes solar development in the state.
Other organizations:
Mississippi Natural Resources, P.O. Box 10600, Jackson, MS 39209
Mississippi Solar Council, 513 N. State St., Jackson, MS 39209

Mississippi Solar Energy Association, 225 W. Lampkin Rd., Starkville, MS 39759

Missouri

Information
Missouri Energy Program
Department of Natural Resources
P.O. Box 1309
Jefferson City, MO 65102
(800) 392-8269 State Energy Hotline
Conducts feasibility studies; distributes basic solar primer and blueprints for active and passive systems. Conducts workshops; offers technical assistance; performs surveys of solar installations and solar education opportunities
Missouri Energy Extension Service,P.O. Box 1309, Jefferson City, MO 65102
Mo-Ark Solar Energy Association, Box 1643, Jefferson City, MO 65102
Solar Research Advisory Panel, c/o Herbert Wade, Department of Natural Resources, Box 1309, Jefferson City, MO 65102

Montana

Information
Renewable Energy Bureau
Department of Natural Resources and Conservation
32 S. Ewing
Helena, MT 59601
(406) 449-4624
Conducts feasibility studies and seminars; administers state renewable energy grant program (includes solar heating, wind energy, geothermal, small-scale hydro and biomass).
Other organizations:
Alternative Energy Resource Organization, 435 Stapleton Bldg., Billings, MT 59101

Nebraska

Information:
Nebraska Solar Office
W-191, Nebraska Hall
University of Nebraska
Lincoln, NE 68588
(402) 472-3414
Conducts workshops; distributes state solar directory, federal solar publications and a general energy newsletter.
Other organizations:
Midwest Energy Alternatives, 3206 S. 127th Street, Omaha, NE 68154
Nebraska State Energy Office, State Capitol, Lincoln, NE 68507
Nebraska Solar Energy Association, University of Nebraska, Dept. of Electrical Technology, 60th & Dodge Streets, Omaha, NE 68182

Nevada

Information
Nevada Department of Energy
Noel A. Clark, Director
400 W. King, Suite 106
Carson City, NV 89710
(702) 885-5157
Activities include several conservation and renewable resource programs involving all energy-consuming sectors. Programs include informational, educational, technical, and referral services. Solar commercialization efforts include distribution of federal and state solar publications, public lectures, workshops, exhibits, and instititutional support programs. Renewable resource directory available soon.
Other organizations:
Nevada Association of Solar Energy Advocates, P.O. Box 8179, University Station, Reno, NV 89507

Nevada Solar Energy Association, 3838 Rayment Drive, Las Vegas, NV 89121

New Hampshire

Information:
Governor's Council On Energy
2 ½ Beacon Street
Concord, NH 03301
(603) 271-2711 Solar Information
(800) 852-3406 State Energy Hotline
Answers consumer questions on solar applications and tax credits, provides training and technical assistance for comercial, residential, and industrial solar.
Other organizations:
New Hampshire Solar Energy Association, P.O. Box 4382, Manchester, NH 03105
Northern New Hampshire Solar Energy Association, c/o EVOG, Hebron, NH 03241
White Mountain Solar Energy Association, 34 Spring Street, Whitefield, NH 03598

New Jersey

Information
Office of Alternative Technology
New Jersey Department of Energy
101 Commerce Street
Newark, NJ 07102
(202) 648-6293 Solar Information
(800) 492-4242 State Energy Hotline
Studies solar and solid-waste recovery energy sources; conducts policy planning, feasiblity studies, seminars and workshops for building inspectors and installers.
Other organizations:
Association of New Jersey Environmental Commissions, P.O. Box 157, Mendham, NJ 07945

New Jersey Energy Extension Service, 101 Commerce Street, Newark, NJ 07102

New Jersey Solar Action, 401 Cooper Street, Camden, NJ 08102

New Jersey Solar Energy Association, 140 Shrewsbury Avenue, Red Bank, NJ 07701

New Mexico

Information

New Mexico Energy & Minerals Department
Energy Conservation & Management Division
P.O.Box 2770
Santa Fe, NM 87501
(505) 827-5621 Solar information
(505) 827-2386 State Energy Extension Service
(800) 432-6782/3 State Energy Hotline
Conducts policy planning, feasibility studies, and workshops; tests collectors; administers state solar research and development program. Administers research-oriented Solar Energy Institute operated by New Mexico State University in cooperation with the Energy & Minerals Department. Energy Extension Service answers technical and non-technical questions, offers information on state solar tax laws, and distributes introductory primers on active, passive, hybrid solar hot water and space heating systems. Contact them for information on short courses for building attached greenhouses.

Other organizations:

Alamogordo Solar Energy Association, 2315 Cherry Lane, Alamagordo, NM 88310

Albuquerque Solar Energy Association, Solar Technology Liaison Division, Sandia Labs 4714-A, Albuquerque, NM 87185

Las Vegas Solar Energy Association, P.O. Box 24, Montezuma, NM 87731

Lee County Solar Energy Association, 1834 Jefferson, Hobbs, NM 88204

New Mexico Energy Extension Service, P.O. Box 2770, Santa Fe, NM 87501
New Mexico Energy Institute, University of New Mexico, Albuquerque, NM 87131
New Mexico Solar Energy Association, P.O. Box 2004, Santa Fe, NM 87501
New Mexico Solar Energy Institute, Box 3 SOL, New Mexico State University, Las Cruces, NM 88003
New Mexico Solar Industry Development, 300 San Mateo, N.E., Suite 805, Albuquerque, NM 87108
Roswell Solar Energy Association, 204 S. Missouri, Roswell, NM 88201
Solar Energy Society of Las Cruces, P.O. Box 1592, Las Cruces, NM 88001
Taos Solar Energy Association, P.O. Box 2334, Taos, NM 87571

New York

Information
New York State Energy Research and Development Agency
Solar Program
Agency Building #2
Empire State Plaza
Albany, NY 12223
(518) 474-8181
Provides consumer information; distributes solar energy and conservation publications: sponsors conferences and workshops; conducts policy planning; assists local governments in promotion of solar energy.
Other organizations:
Eastern New York Solar Energy Association, P.O. Box 5181, Albany, NY 12205
Metropolitan New York Solar Energy Society, P.O. Box 2147, Grand Central Station, New York, NY 10163
Mid-Hudson Renewable Energy Association, PO Box 909, Millbrook, NY 12545

New York Energy Extension Service, 2 Rockefeller Plaza, Albany, NY 12223
Northeast Solar Energy Center, 470 Atlantic Avenue, Boston, MA 02140
Solar Energy Association of Western New York,P.O. Box 501, Belfast, NY 14711
Solar Utilization in Northwest New York, P.O. Box 9501, Rochester, NY 04604
Southern Tier Solar Energy Association, P.O. Box 3004, Elmira, NY 14905

North Carolina

Information
Information Section
Energy Division
North Carolina Department of Commerce
P.O. Box 25249
Raleigh, NC 27611
(800) 662-7131 State Energy Hotline
Conducts policy planning; funds a solar technical assistance program for speculative home builders; funds solar training programs for vocational and technical schools; provides individuals with the latest information on state and federal energy programs and policies, alternative energy sources and energy legislation; and conducts workshops. Distributes a state solar directory and federal solar and other energy publications.
Other organizations:
North Carolina Coalition for Renewable Energy Resources, P.O. Box 10564, Raleigh, NC 27605
North Carolina Energy Extension Service, Department of Commerce, Energy Division, P.O. Box 25249, Raleigh, NC 27611
North Carolina Land Trust Association, 719 Ninth St, Suite 206, Durham, NC 27705
North Carolina Solar Energy Association 7001 Buckhead Drive, Raleigh, NC 27609

North Dakota

Information

Energy Information
Office of Energy Management & Conservation
1533 N. 12th Street
Bismarck, ND 58501
(701) 224-2250
Conducts workshops, answers questions on energy (including solar); schedules speaking engagements.

Other organizations:

Dakota Resource Council, P.O. Box 254, Dickison, ND 58601
Energy Association of North Dakota, P.O. Box 8103, University Station, Grand Forks, ND 58202
North Dakota Energy Extension Service, 1533 N. 12th Street, Bismarck, ND 58501

Ohio

Information

Ohio Department of Energy
34th Floor
30 E. Broad Stret
Columbus, OH 43215
(614) 466-8277
(800) 282-9284
Conducts policy planning, seminars and feasibility studies of solar industrial process heat in cooperation with Ohio State University. Publishes "Energy-Saving Home Improvement Guide," available through toll-free number.

Other organizations:

Columbus Solar Energy Society, P.O. Box 333, Dublin, OH 43017
Ohio Energy Extension Service, 30 E. Broad Street, Columbus, OH 43215
Ohio Solar Energy Association, Wright State University, Dayton, OH 45435
Ohioans of Utility Reform, 842 S. Green Rd., Cleveland, OH 44121

Oklahoma

Information

Energy Information
Oklahoma Department of Energy
4400 N. Lincoln Blvd.
Suite 251
Oklahoma City, OK 73105
(405) 521-3941
Conducts policy planning and feasibility studies; answers solar and general energy-related questions.

Other organizations:

Institute for Energy Analysis, Oklahoma State University, Stillwater, OK 74074
Oklahoma Energy Extension Service, 4400 N. Lincoln Blvd., Suite 251, Oklahoma City, OK 73105
Oklahoma Environmental Information & Media Center, East Central University, Ada, OK 74820
Oklahoma Solar Energy Association, Solar Energy Laboratory, University of Tulsa, Tulsa, OK 74104
Oklahoma Solar Industries Association, 8213 N. Classen, Oklahoma City, OK 73114

Oregon

Information

Oregon Department of Energy
Labor & Industries Bldg.
Salem, OR 97310
(800) 452-7813 "Access 800" State Government Hotline
(503) 378-4040
Conducts policy planning, feasibility studies, and workshops; certifies active and passive systems for state solar tax credit; answers questions referred from "Access 800."

Other organizations:

Columbia Solar Energy Association, 1018 N. Ainsworth, Portland, OR 97221
Eastern Oregon Solar Group, c/o Community Schools, 410 S.W. 13th, Pendleton, OR 98701

Oregon Energy Extension Service, Oregon State University, Corvallis, OR 97531

OSMI Energy Center, 4015 S.W. Canyon Rd., Portland, OR 97221

Oregon Solar Energy Group, Box 751, Portland, OR 97207

Oregon Solar Energy Industries Association, 7645 S.W. Capitol Highway, Suite B, Portland, OR 97225

Oregon Solar Energy Society, P.O. Box 142, Salem, OR 97308

Oregon Solar Institute, 637 S.E. Harrison Street, Portland, OR 97214

Responsible Urban Neighborhood Technology, 2926 N. Williams, Portland, OR 97227

Solar Energy Resource Group, 457 Ponderosa Drive, Roseburg, OR 97470

Pennsylvania

Information

Governor's Energy Council
1625 Front Street
Harrisburg, PA 17102
(800) 882-8400
(717) 783-8610
Conducts market research; inventories state energy resources; studies feasibility of solar retrofits for public buildings; conducts workshps. Developing data base for alternative energy sources for Pennsylvania.

Other organizations:

Bucks County Solar Energy Association, 305 9th Street, Sellersville, PA 18960

Grassroots Alliance for a Solar Pennsylvania, Community Education Center, 3500 Lancaster Avenue, Philadelphia, PA 19104

Lehigh Valley Solar Energy Association, P.O. Box 253, Lehigh Valley, PA 18001

Pennsylvania Solar Energy Industries Association, 1930 Brandon Road, Norristown, PA 19403

Pennsylvania Solar Power Advocates, 615 Hedgerow Lane, Lancaster, PA 17601
Philadeplphia Solar Energy Association, 2233 Grays Ferry Avenue, Philadelphia, PA 19146
South Central Pennsylvania Solar Energy Association, P.O. Box 1378, State College, PA 16801
Western Pennsylvania Solar Energy Association, P.O. Box 81061, Pittsburgh, PA 15217

Puerto Rico

Information
Puerto Rico Office of Energy
Energy Information Program
P.O. Box 41089, Minillas Station
Santurce, PR 00940
(800) 727-8877 State Energy Hot-line
Maintains energy information center. Develops and distributes energy-related educational materials. Answers solar and energy related questions. Offers state directory of solar water heaters distributors.
Other organizations:
Puerto Rico Energy Extension Service, P.O. Box 41089, Minillas Station, Santurce, PR 00940
Puerto Rico Solar Energy Association, P.O. Box 3011, San Juan, PR 00936

Rhode Island

Information
Governor's Energy Office
80 Dean Street
Providence, RI 02903
(401) 277-3370
Conducts policy planning and coordination, solar economic studies, energy conservation training and workshops. Maintains energy information center; answers technical and nontechnical questions. Publishes Southern New England Solar

Directory, list of solar buildings in Rhode Island, solar and alternative energies handbook, and other publications.

Other organizations:

Rhode Island Energy Extension Service, 80 Dean Street, Providence, RI 02903

Rhode Island Solar Energy Association, P.O. Box 212, Providence, RI 02193

South Carolina

Information

South Carolina Energy Management Office
SCN Center, Suite 1130
1122 Lady Street
Columbia, SC 29201
(803) 758-2050
(800) 922-1600

Develops energy conservation courses for vocational schools; conducts seminars and policy planning.

Other organizations:

South Carolina Energy Extension Service, South Carolina State Board for Technical and Comprehensive Education, 1429 Senate Street, Columbia, SC 29201

South Dakota

Information

Office of Energy Policy
Capitol Lake Plaza Bldg.
Pierre, SD 57501
(800) 592-1865

Collects and disseminates solar information. Participates and assists in development, demonstration, education and policy recommendations necesary to aid in the commercialization effort of renewable energy resources within South Dakota.

Other organizations:

Governor's Advisor on Solar Energy, Box 1311, Rapid City, SD 57701

South Dakota Energy Extension Service, Office of Energy Policy, Capitol Lake Plaza Bldg., Pierre, SD 57501
South Dakota Solar Energy Association, c/o Steve Wegman, Part A, Pierre, SD 57501
South Dakota Renewable Energy Association, P.O. Box 782, Pierre, SD 57501

Tennessee

Information
Tennessee Energy Authority
276 Capitol Boulevard Bldg.
Suite 707
Nashville, TN 37219
(651) 741-6671 Solar Information
(800) 342-1340 Energy Hotline
Provides information and advice to parties interested in active or passive technolgy. Studies existing state solar buildings: conducts feasiblity studies, organizes workshops on passive solar design for builders, developers, and engineers, Offers information on the TVA Solar Hot Water Demonstration Program.
Other organizations:
Tennessee Energy Extension Service, Tennessee Energy Authority, 226 Capitol Blvd. Bldg., Suite 707, Nashville, TN 37219
Tennessee Environmental Council, P.O. Box 1422, Nashville, TN 37202
Tennessee Solar Energy Association, P.O. Box 448, Jefferson City, TN 37760

Texas

Information
Texas Energy and Natural Resources Advisory Council
200 E. 18th Street
Austin, TX 78701

Alternative energy research and development; conducts policy planning, feasibility studies, Distributes research program reports, quarterly newsletter on state Energy Advisory Council activities, and other publications. Ask for their publications list.
Other organizations:
Texas Energy Extension Service, Texas A&M University, College Station, TX 78843
Texas Solar Energy Society, (TX-SES), 1007 S. Congress Ave., Suite 348, Austin, TX 78704

Utah

Information
Utah Energy Office
825 North 3rd West
Salt Lake City, UT 84103
(801) 533-3228
(800) 662-3663
Answers solar and general energy related questions. Gives solar energy presentations to elementary students; has a solar energy library.
Other organizations:
Utah Energy Extension Office, 825 North 3rd West, Salt Lake City, UT
Utah Solar Advocates, P.O. Box 1405, Salt Lake City, UT 84110
Utah Solar Energy Society, P.O. Box 6032, Salt Lake City, UT 84106

Vermont

Information
Vermont Energy Office
State Office Bldg.
Montpelier, VT 05602
(800) 642-3281 State Energy Action Line
Answers questions and offers solar advice to consumers; studies feasiblity of solar energy retrofits; conducts solar seminars and

workshops.
Other organizations:
Solar Association of Vermont, 73 Main Street, Montpelier, VT 05602

Virginia

Information
Office of Emergency and Energy Services
310 Turner Road
Richmond, VA 23235
Maintains solar energy library; answers technical and non technical questions. Distributes fact sheets, bibliographies, and state solar directory.
Other organizations:
Roanoke Valley Solar Association, 312 First St., SE, Roanoke, VA 24011
Virginia Energy Extension Service, 310 Turner Road, Richmond, VA 23235
Virginia Renewable Energy Lobby, 203 North Main Street, Lexington, VA 24450
Virginia Solar Council, 2338 N. 11th St., Rm. 302, Arlington, VA 2201
Virginia Solar Energy Association, c/o Piedmont Technical Associates, 300 Lansing Ave., Lynchburg, VA 24503.

Washington

Information
State Energy Office
40 E. Union Street
First Floor
Olympia, WA 98504
(206) 754-0731 Solar Information
(206) 344-3440 State Energy Extension Service
Conducts policy planning, feasibility studies, regional conferences and workshops. Factsheets, bibliographies, and do-it yourself plans and an extensive energy library.

Other organizations:

Citizens for a Solar Washington, P.O. Box 20123, Broadway Station Seattle, WA 98102

Ecotope Group, 2332 E. Madison, Seattle, WA 98112

Inland Empire Solar Energy Association, P.O. Box 13, Spokane, WA 99210

Kittias Valley Alternative Energy Association, P.O. Box 282, Ellensburg, WA 98926

Klickitat Citizens for Energy Awareness, P.O. Box 115, White Salmon, WA 98672

Methow Solar Energy Association, Twisp, WA 98856

Northeast Washington Appropriate & Creative Technology, Rte. 1, Box 149, Republic, WA 99168

Pacific Northwest Solar Energy Association, Puget Power Bldg., Bellevue, WA 98008

Pond Oreille Center for Appropriate Technology, Rte. 2, Box 770, Newport, WA 99156

Ryegrass Energy Resources Group, 611 E. Rose, Walla Walla, WA 99362

Snohomish County Solar Energy Association, 425 Union Ave., #B, Snohomish, WA 98290

Solar Wentachee, 620 Lewis St., Wentachee, WA 98801

Southern Puget Sound Solar Energy Assocation, P.O. Box 454, Olympia, WA 98506

Tri-Cities Solar Energy Society, 1732 W. Irving, Pasco, WA 99301

Washington Energy Extension Service, Cooperative Extension, Washington State University, Pulman, WA 99164

Western Washington Solar Energy Association, 314 Economy Bldg., 93 Pike Street, Seattle, WA 98101

Whatcom Solar Energy Association, P.O. Box 2035, Bellingham, WA 98227

Whidbey Island Solar Energy Assocation, 2920 E. High Crest Road, Langley, WA 98260

West Virginia

Information

Solar Energy
Fuel & Energy Office
1591 Washington St. E
Charleston, WV 25311
(800) 642-9012-3 State Energy Hotline
Offers state directory of solar architects and engineers; distributes federal solar publications.

Other organizations:

West Virginia Energy Extension Service, 1591 Washington Street East, Charleston, WV 25311

Wisconsin

Information

Division of Energy
State Office Bldg.
8th Floor
101 S. Webster Street
Madison, WI 53702
(608) 266-9861
Conducts policy planning, data surveys, state solar installations survey, legislation development, and distributes solar energy packet.

Other organizations:

Wisconsin Solar Energy Association, 1131 University Avenue, Madison, WI 53715

Wyoming

Information

Wyoming Energy Conservation Office
320 W. 25th Street
Capitol Hill Bldg.
Cheyenne, WY 82001
(307) 777-7131
Formulates and conducts state policy and planning regarding

energy conservation. Develops, reviews and disseminates information on audit materials, educational literature, technical and community grants and renewable energy sources. Develops and promotes building code legislation, industrial and commercial conservation programs and consumer awareness workshops. Provides general information and assistance to public and private sectors.

Other organizations:

Community Solar Greenhouse, Community Action of Laramie County Inc., 1603 Central Avenue, Suite 400, Cheyenne, WY 82001

Solar Advisory Group, c/o RMIEE, Box 3965, University of Wyoming, Laramie, WY 82071

Wyoming Energy Extension Service, P.O. Box 3965, University Station, Laramie, WY 82071

Wyoming Solar Alliance, P.O. Box 12, Hillsdale, WY 82060

Regional groups

Northeast

Mid-Atlantic Solar Energy Association
, 2233 Gray's Ferry Ave., Philadelphia, PA 19146
New England Solar Energy Association, PO Box 541, 22 High St., Brattleboro, VT 05301
Northeast Energy Center, 470 Atlantic Ave., Boston, MA 02110

South

Southern Unity Network/Renewable Energy Projects, 3110 Maple Dr., Atlanta, GA 30305
Tennessee Valley Authority, 400 Commerce Ave., Knoxville, TN 37902

Middlewest

Midland Energy Institute, 906 Grand, Suite 100, Kansas City, MO 64106

Sun/Tronic house designed by Barry A. Berkno, built by the Copper Development Association, includes active and passive solar, and photovoltaics.

Solar Houses Around the Country

This section of *The Solar Energy Almanac* presents new passive solar houses that have been built all over the United States - from Maine to Arizona to Alaska to Georgia. These houses were all built with partial funding by the Department of Housing and Urban Development, and they are proof that solar works virtually everywhere in the country.

HUD's main purpose in funding these houses, aside from further demonstration of the viability of passive solar heating, was to show that solar could be easily adapted to the tastes of the mainstream of American home buyers. They wanted to prove that a solar home didn't have to look like something strictly for the space age.

It is generally believed that passive solar heating increases the cost of a home by from 5 to 10 percent.

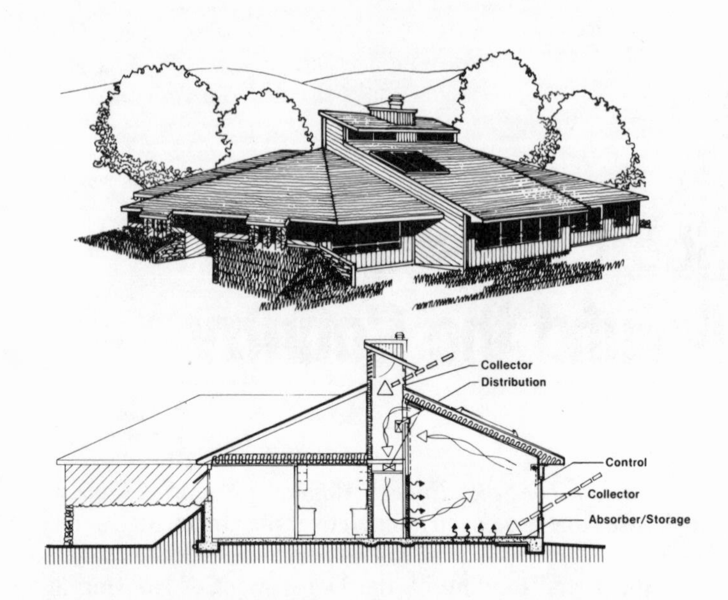

Prattville, Alabama

Builder: Simmons Builders, Prattville, AL
Designer: Chambless-Killingsworth & Associates, Montgomery, AL
Price: $80,000
Heated area: 2,066 square feet
Heat load: 38 million Btus per year
2,291 Degree Days
Solar heat: 37 percent
Direct gain, sun-tempering
Collector: South-facing double-glazed window, 169 square feet; *Absorber*: Floor quarrry tile; *Storage*: Concrete floor slab-capacity: 11,650 Btus; *Distribution*: Natural and forced convection, radiation; *Controls*: Insulating shutters
Back-up: Gas-fired hot air furnace
Domestic hot water: 3 flat-plate collectors (58 square feet), 66 gallon storage
Cooling: Natural ventilation

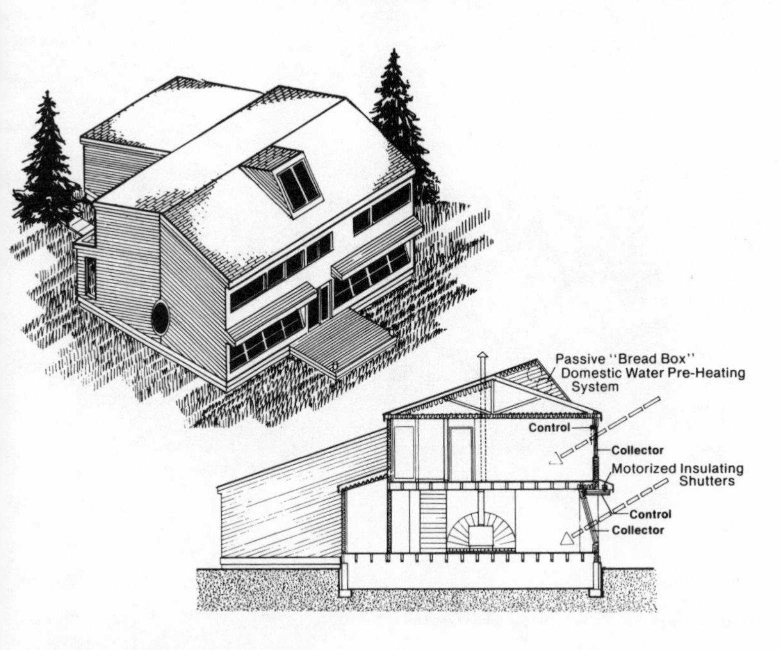

Eagle River, Alaska

Builder: Land Trust Realty/Development Conservation Enterprises, Eagle River, AK
Designer: Architecture/Environmental Design, Anchorage, AK
Solar designer: Jimmy Anderson
Price: $122,000
Heated area: 2,219 square feet
Heat load: 102 million Btus per year
10,850 Degree Days
Solar heat: 42 percent
Sun-tempering, direct gain
Collector: South-facing glazing, 313 square feet; *Absorber:* Living room, dining room, bedrooms; *Storage:* Same as absorber-capacity: 3,997 Btus; *Distribution:* Radiation; *Controls*:Exterior shutters, overhang, movable insulation
Back-up: Gas furnace (32,000 Btus per hour), woodstove
Domestic hot water: Passive preheat (36 square feet)

Mat-Su Borough, Alaska

Builder: Solarctic Construction Company, Anchorage, AK
Designer: Ronald J. Bissett, Anchorage, AK
Price: $85,000
Heated area: 1,466 square feet
Heat load: 58 million Btus per year
10,667 Degree Days
Solar heat: 62%
Direct gain, indirect gain (Trombe wall), isolated gain
Collector: Single-glazed Trombe wall glass, double-glazed greenhouse glass, triple-glazed glass, 565 square feet; *Absorber*: Tile pavers, greenhouse floor, concrete block Trombe wall; *Storage*: Concrete slab flor, Trombe wall, greenhouse floor (capacity: 19,264 Btus); *Distribution*: Radiation, natural convection; *Controls*: Retractable shades, Trombe wall registers, floor vents, fans, movable insulation
Back-up: Woodburning stove (50,000 Btus per hour); electric resistance heaters (25,000 Btus per hour)
Domestic hot water: Liquid flat-plate collectors (129 square feet), 120-gallon storage

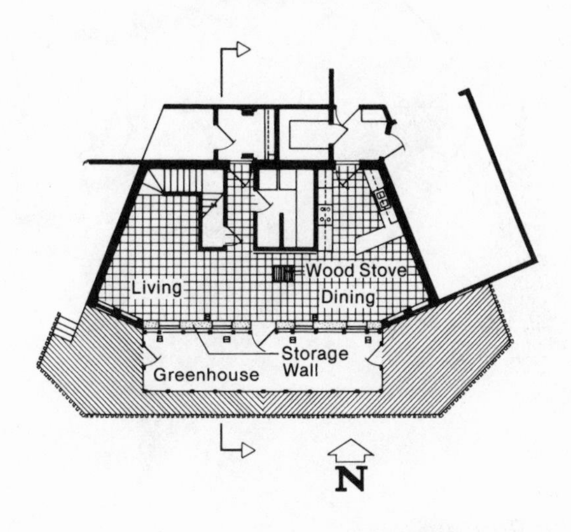
Wood Stove
Living
Dining
Storage
Wall
Greenhouse
N

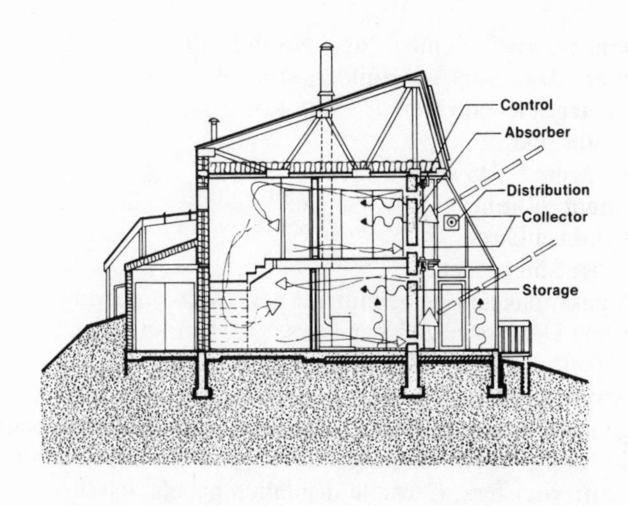
Control
Absorber
Distribution
Collector
Storage

North Little Rock, Arkansas

Builder: Winrock Homes, Inc., North Little Rock, AR
Designer: Arkansas Ark Builders, Inc., Little Rock, AR
Solar designer: Bob Bland, Little Rock, AR
Price: $68,900
Heated area: 1,515 square feet
Heat load: 24 million Btus per year
3,219 Degree Days
Solar heat: 58 percent
Direct gain, isolated gain, indirect gain, sun-tempering
Collector: Double-glazed windows, greenhouse glass, 372 square feet; *Absorber*: Black-painted metal sheets in thermosiphon ducts, ceramic tile floor; *Storage*: Concrete storage slab, concrete block walls, rock storage bin (capacity: 26,856 Btus); *Distribution*: Radiation, convection; *Controls*: Baseboard registers, movable insulation panels, dampers, overhang;
Back-up: Woodburning fireplace, gas furnace (40,000 Btus per hour)

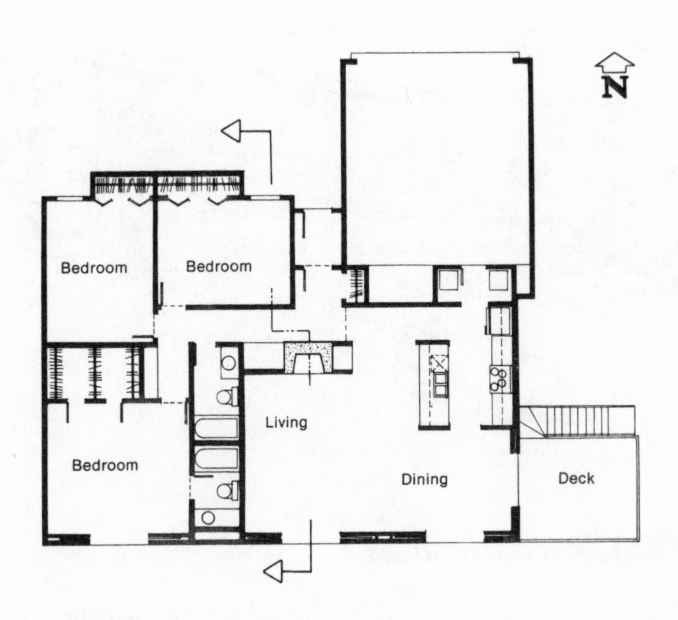
N
Bedroom
Bedroom
Living
Bedroom
Dining
Deck

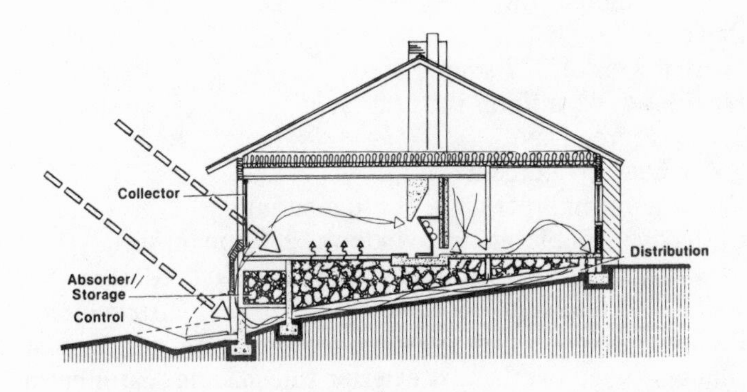
Collector
Absorber//
Storage
Control
Distribution

West Sedona, Arizona

Builder: Gary E. Wagley, General Contractor, Sedona, AZ
Designer: Sun System Engineering, West Sedona, AZ
Solar designer: Jim Faney, Sedona, AZ
Price: $93,000
Heated area: 1,377 square feet
Heat load: 45 million Btus per year
3,702 Degree Days
Solar heat: 84 percent
Direct gain, indirect gain, sun-tempering
Collector: Double-glazed windows, 280 square feet; *Absorber*: Tile floors and mass walls; *Storage*: Masonry walls, concrete slab (capacity: 46,074 Btus); *Distribution*: Radiation, natural and forced convection; *Controls*: Insulated shutters, fixed overhangs, exterior sun shading, earth berms, clerestory vent
Back-up: Electric furnace and woodburning stove
Domestic hot water: Two roof-mounted flat-plate collectors, 82 gallon glass-lined tank
Cooling: Natural and induced ventilation, night-sky radiation

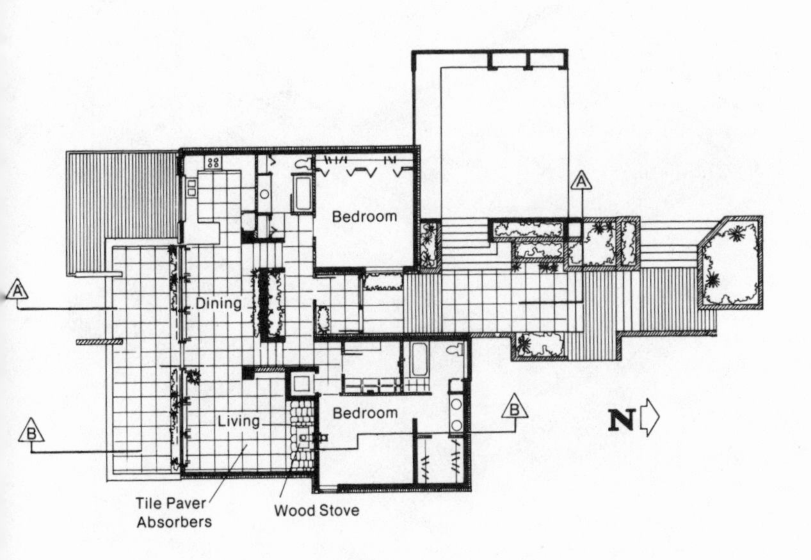
A
Bedroom
A
Dining
Living
Bedroom
B
B
N
Tile Paver Absorbers
Wood Stove

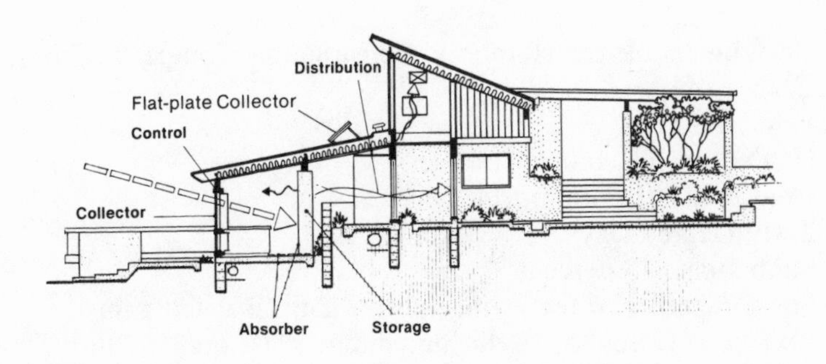
Distribution
Flat-plate Collector
Control
Collector
Absorber
Storage

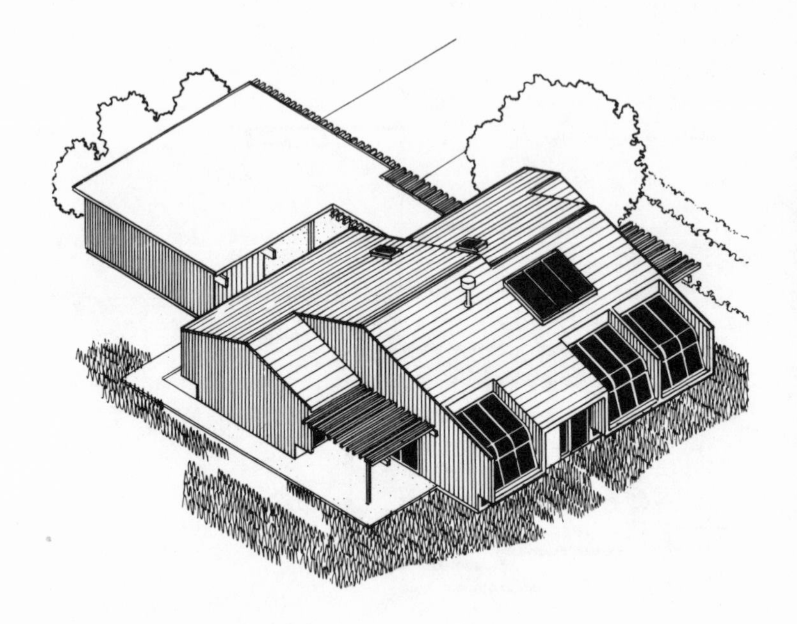

Davis, California

Designer/Builder: Hornbeek Construction Company, Davis, CA
Price: $90,000
Heated area: 1,180 square feet
Heat load: 41 million Btus per year
2,419 Degree Days
Solar heat: 75 percent
Sun-tempering, direct gain, indirect gain, isolated gain
Collector: Greenhouse glazing, sliding glass doors, 286 square feet; *Absorber*: Greenhouse brick floor over concrete slab and sand, steel water tank surface; *Storage*: Brick floor, concrete slab floor, water in steel tanks (capacity: 8,906 Btus); *Distribution*: Radiation, natural ventilation; *Controls*: Insulated curtain, greenhouse canvas shade, adjustable sunshade, tank insulating curtain, vents.

Active solar heating: Active space heating back-up from domestic hot water system
Back-up: Hot water is pumped through fan-coil units and baseboard
Cooling: Night-sky radiation, convection

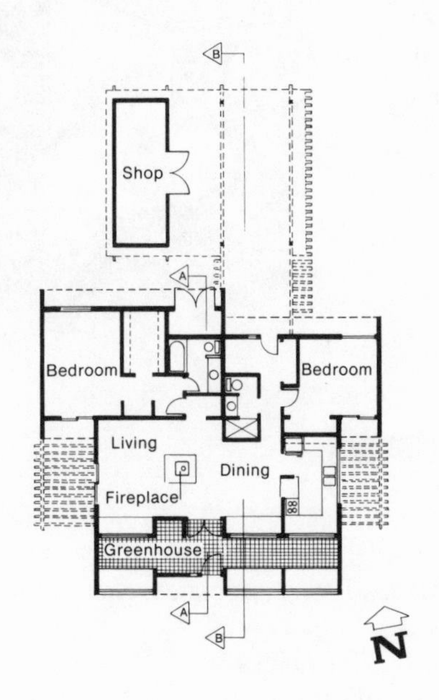

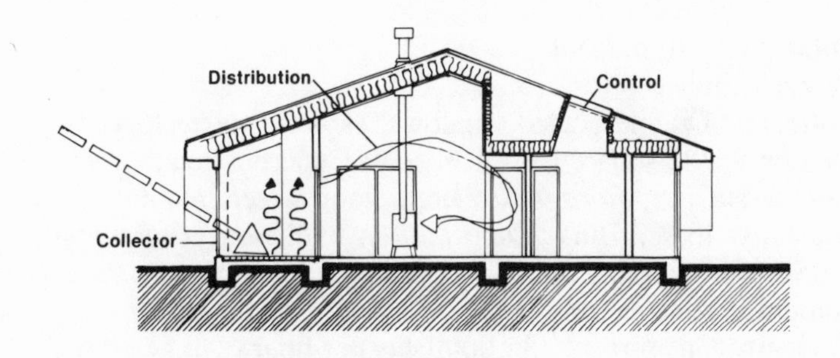

Davis, California

Builder: Walker, Donat & Co/Lien, Stonegate Company, Inc., Sacramento, CA
Designer: Charles Eley Associates, San Francisco, CA
Price: $50,000
Heated area: 1,273 square feet
Heat load: 40 million Btus per year
2,819 Degree Days
Solar heat: 70 percent
Direct gain
Collector: Double-glazed windows, skylight, clerestory windows, 282 square feet; *Absorber*: Concrete floor, water column surface; *Storage*: Concrete floor, water column (capacity: 13,122 Btus); *Distribution*: Radiation, natural and forced convection; *Controls*: Overhang, ducts, vents, draperies, sunscreens
Back-up: Gas furnace (36,000 Btus per hour)
Domestic hot water: Active, liquid flat-plate collectors (42 square feet), 82-gallon storage

Bedroom
Bedroom
Bedroom
Dining
Living
N

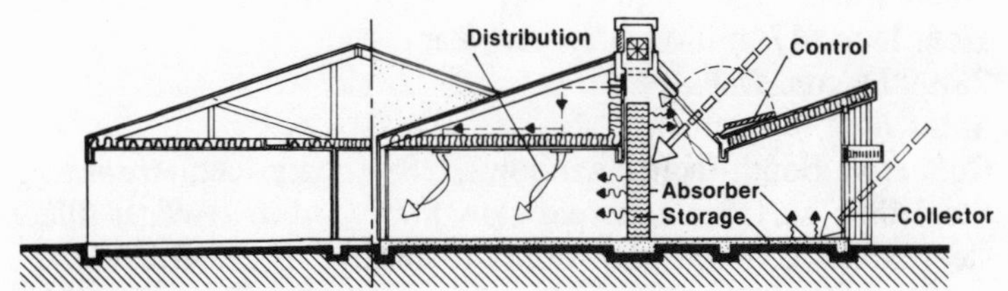
Distribution
Control
Absorber
Storage
Collector

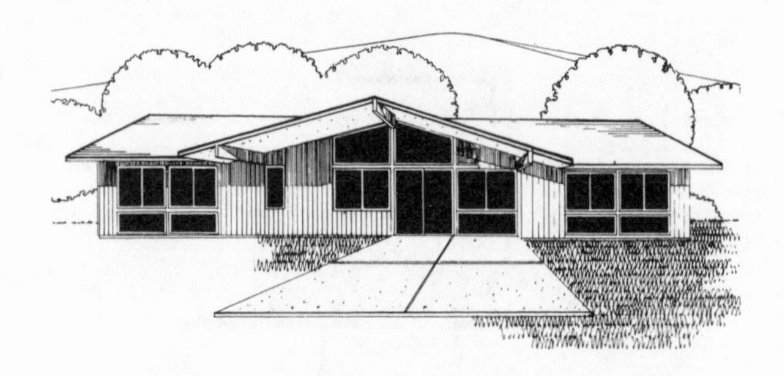

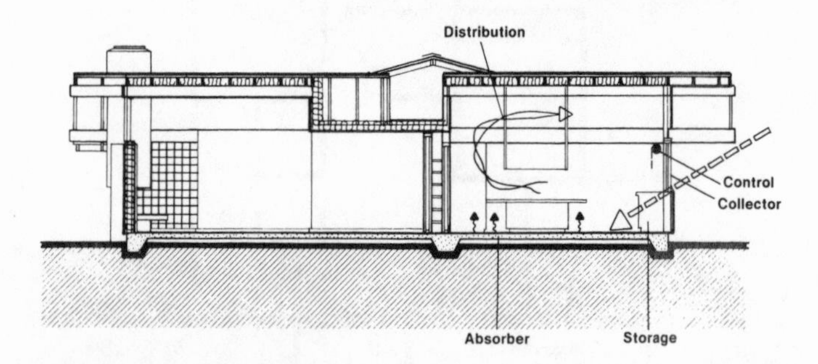

Elk Grove, California

Builder: Streng Brothers Homes, Inc., Davis, CA
Designer: Carter Sparks, James Plumb, Sacramento, CA
Solar designer: James Plumb
Price: $83,000
Heated area: 1,886 square feet
Heat load: 47 million Btus per year
2,499 Degree Days
Solar heat: 64 percent
Collector: South-facing windows, 288 square feet; *Absorber*: Steel tank surface, concrete slab floor; *Storage*: Water-filled steel tanks, concrete slab floors (capacity: 13,122 Btus); *Distribution*: Radiation, natural and forced convection; *Controls*: Insulating drapes, thermostat
Back-up: Gas furnace (36,000 Btus per hour)
Cooling: Natural and induced ventilation

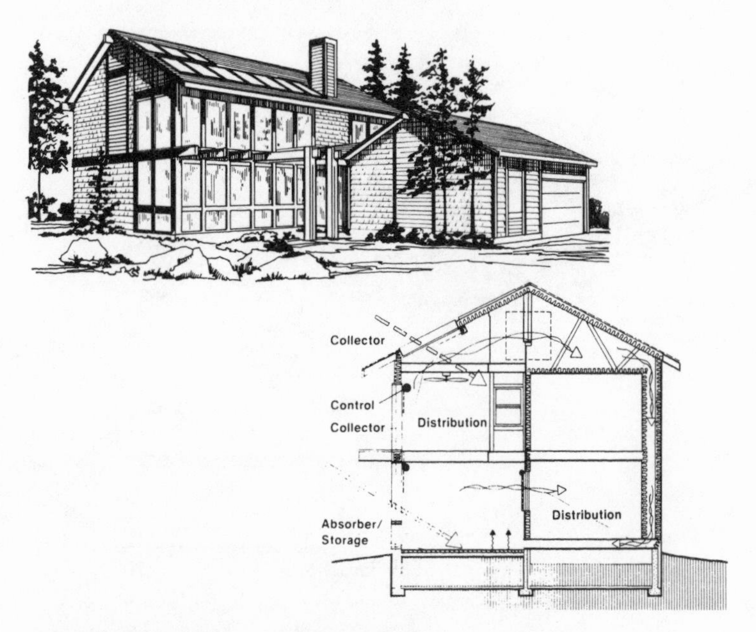

Colorado Springs, Colorado

Designer/Builder: Rick A. Cowlishaw, Colorado Homes, Inc., Colorado Springs, CO
Price: $85,000
Heated area: 1,840 square feet
Heat load: 38 million Btus per year
6,432 Degree Days
Solar heat: 82 percent
Direct gain, isolated gain
Collector: Greenhouse glazing, south-facing windows, 658 square feet; *Absorber*: Brick pavers over concrete floor, surface of concrete walls, hot tub; *Storage*: Concrete floor and walls, hot tub (capacity: 96,000 Btus); *Distribution*: Radiation, natural and forced convection; *Controls*: Vents, blinds, overhangs
Back-up: Electric resistance baseboard heaters
Domestic hot water: Passive "bread-box" preheat, 40-gallon storage

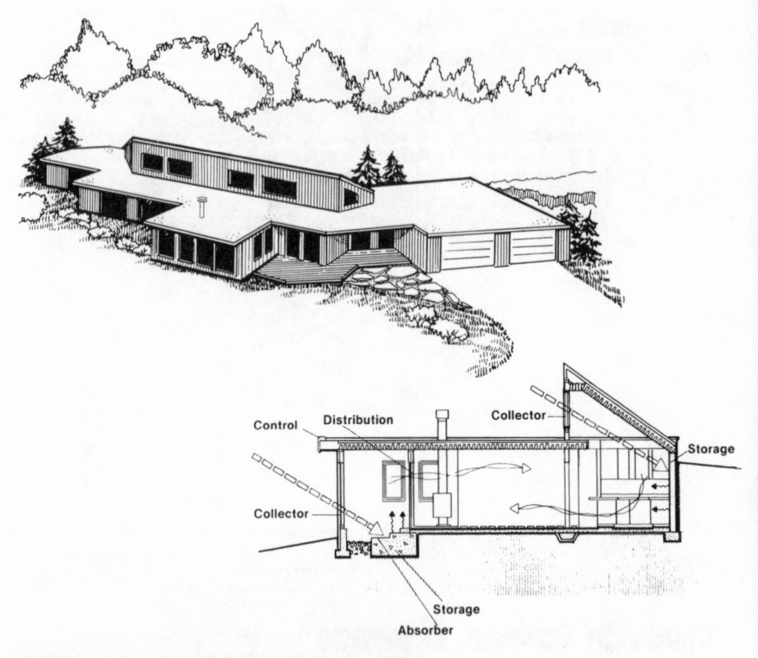

Colorado Springs, Colorado

Builder: Energy Techniques, Colorado Springs, CO
Designer: Design Group Architects, Colorado Springs, CO
Price: $85,000
Heated area: 1,558 square feet
Heat load: 75 million Btus per year
6,423 Degree Days
Solar heat: 48 percent
Direct gain, sun-tempering, isolated gain
Collector: Double-glazed windows and glass sliding doors, triple-glazed clerestory windows, 309 square feet; *Absorber*: Brick pavers over concrete slab floor, concrete walls; *Storage*: Concrete slab floor, concrete walls (capacity: 8,804 Btus); *Distribution*: Natural and forced convection, radiation; *Controls*: Thermostatically controlled fans, vents
Back-up: Natural gas forced-air furnace (39,500 Btus per hour), airtight woodburning stove

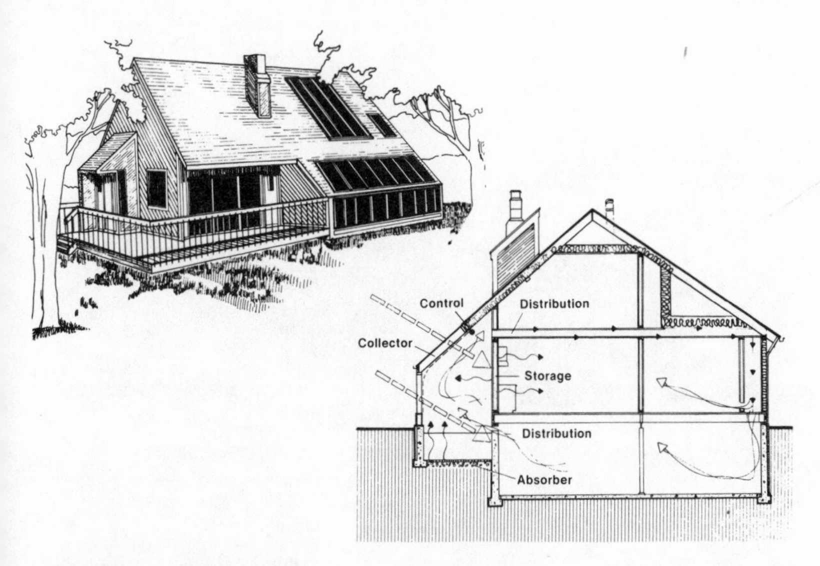

Walden, Colorado

Builder: Habitat Design and Construction Company, Walden, CO
Designer: Passive Solar Home Design Company, Denver, CO
Price: $75,000
Heated area: 1,660 square feet
Heat load: 168 million Btus per year
10,426 Degree Days
Solar heat: 36 percent
Direct gain, isolated gain, sun-tempering
Collector: Greenhouse glazing, south-facing glass doors, 295 square feet; *Absorber*: Brick floor and wall, concrete wall, clay-tiled floor, water-filled benches; *Storage*: Brick floor and wall, concrete wall, clay-tiled floor, water filled benches (capacity: 10,547 Btus); *Distribution*: Radiation, natural and forced convection; *Controls:* Thermostat, ducts, vents, louvers, shades
Back-up: Gas boiler (100,000 Btus per hour)
Domestic hot water: Liquid flat-plate collectors (70 square feet), 82-gallon storage

Carbonadale, Colorado

Duilder: West Sopris Creek Builders
Designer: Sunshine Design, Carbondale, CO
Price: $97,500
Heated area: 1,646 square feet
Heat load: 48 million Btus per year
7,339 Degree Days
Solar heat: 56 percent
Direct gain, indirect gain (Trombe wall), isolated gain
Collector: South-facing glazing, greenhouse glazing, skylight, 348 square feet; *Absorber*: Concrete floor slab, Trombe wall surface; *Storage:* Concrete floor slab, concrete Trombe wall, mass hearth (capacity: 12,870 Btus); *Distribution*: Radiation, natural and forced convection; *Controls*: Trombe wall shade, insulated shutters and curtains, vents, roll-down shades
Back-up: Electric resistance heater (32,000 Btus per hour), woodburning stove
Domestic hot water: 46 square feet flat-plate collectors, 82-gallon storage

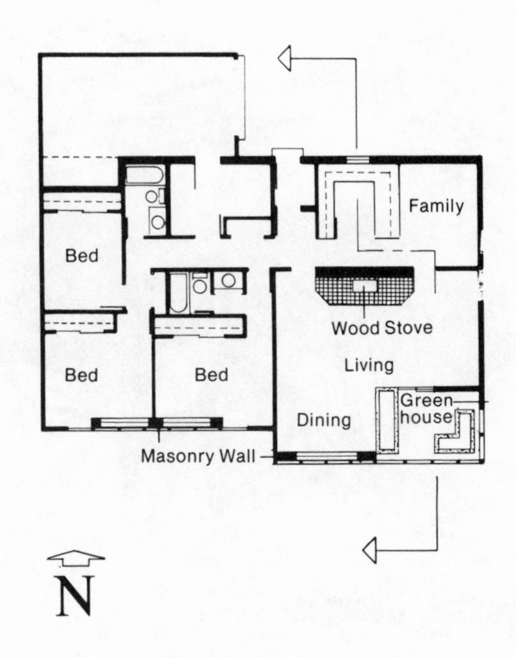
Family
Bed
Wood Stove
Living
Bed
Bed
Green-
house
Dining
Masonry Wall
N

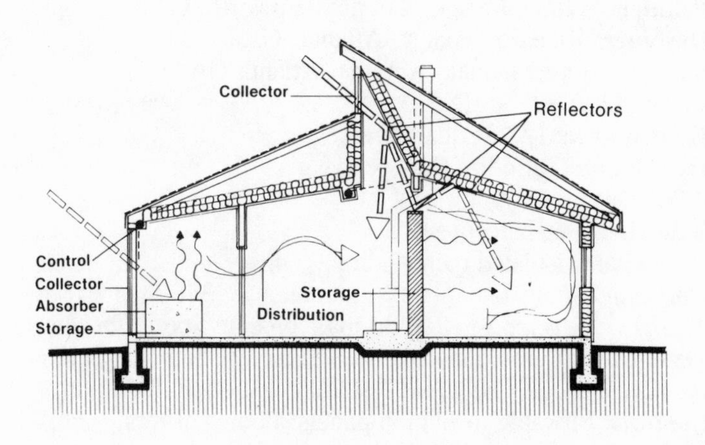
Collector
Reflectors
Control
Collector
Absorber
Storage
Storage
Distribution

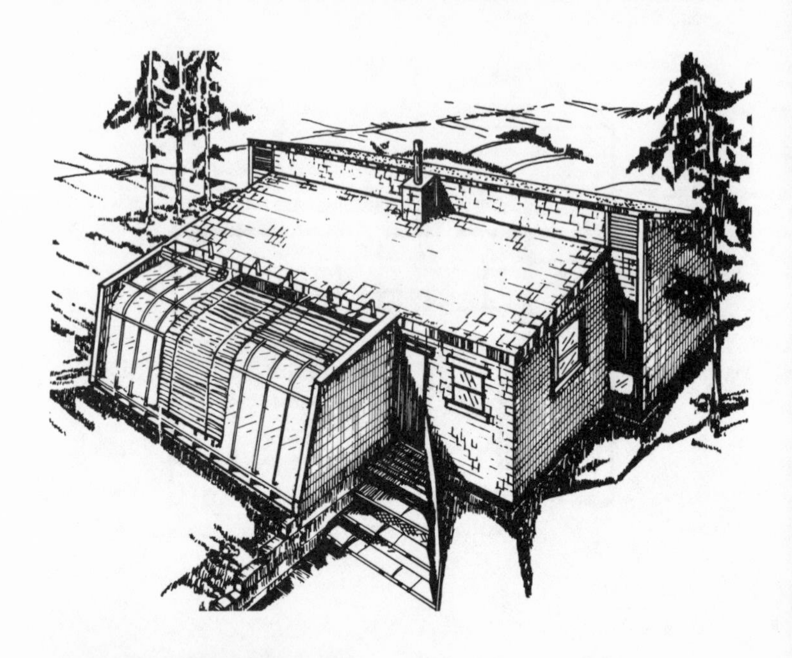

McCaysville, Georgia

Builder: Walnut Town & Country, Epworth, GA
Designer: Richard Seedorf, Atlanta, GA
Solar designer: Donald Abrams, Atlanta GA
Price: $29,500
Heated area: 1,073 square feet
Heat load: 27 million Btus per year
2,961 Degree Days
Solar heat: 81 percent
Direct gain, isolated gain, indirect gain
Collector: South-facing glass, greenhouse, 376 square feet; *Absorber*: Black 55-gallon drums, concrete floor; *Storage*: Water-filled drums, masonry wall, concrete floor (capacity: 16,485 Btus); *Distribution*: Natural convection, radiation; *Controls*: Movable insulation panels, floor registers, backdraft dampers, shades
Back-up: Electric resistance heaters; woodburning stove
Cooling: Induced ventilation

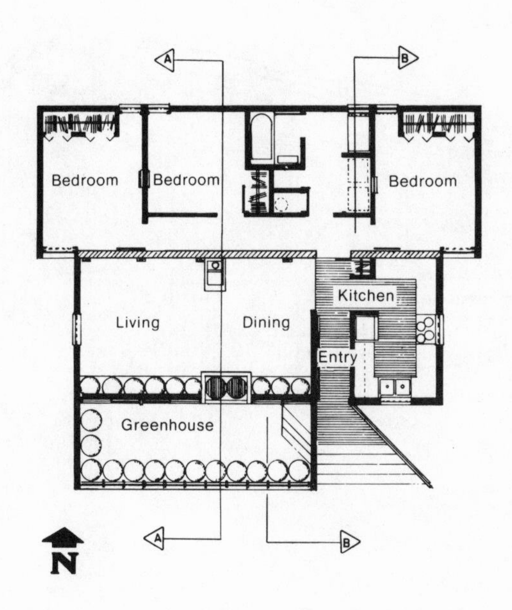
A
B
Bedroom
Bedroom
Bedroom
Kitchen
Living
Dining
Entry
Greenhouse
A
B
N

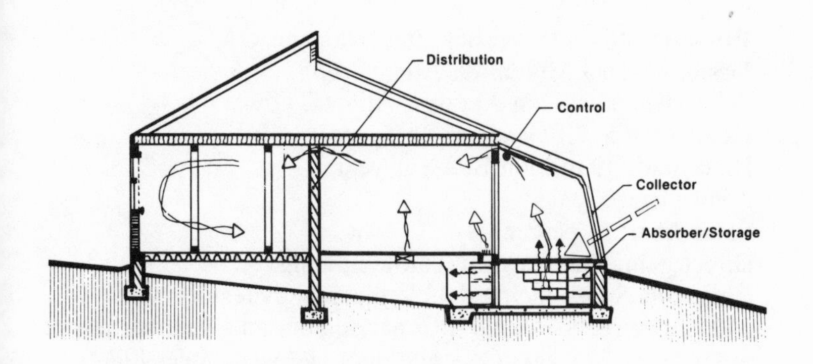
Distribution
Control
Collector
Absorber/Storage

Norcross Gwinnett, Georgia

Builder: Millen Properties, Inc., Atlanta, GA
Designer: Paul Muldawer, Atlanta, GA
Solar designer: Don Abrams, Atlanta, GA
Heated area: 2,039 square feet
Heat load: 49 million Btus per year
2,961 Degree Days
Solar heat: 51 percent
Direct gain, indirect gain, sun-tempering
Collector: South-facing double glazing, clerestory windows, 342 square feet; *Absorber*: Concrete floor and wall, surface of water tubes; *Storage*: Concrete floor and wall, water filled tubes (capacity: 35,324 Btus); *Distribution*: Radiation, natural convection; *Controls*: Overhangs, shading panels, water-tube covers, shading louvers
Back-up: Gas furnace (40,000 Btus per hour)
Domestic hot water: 80-gallon storage

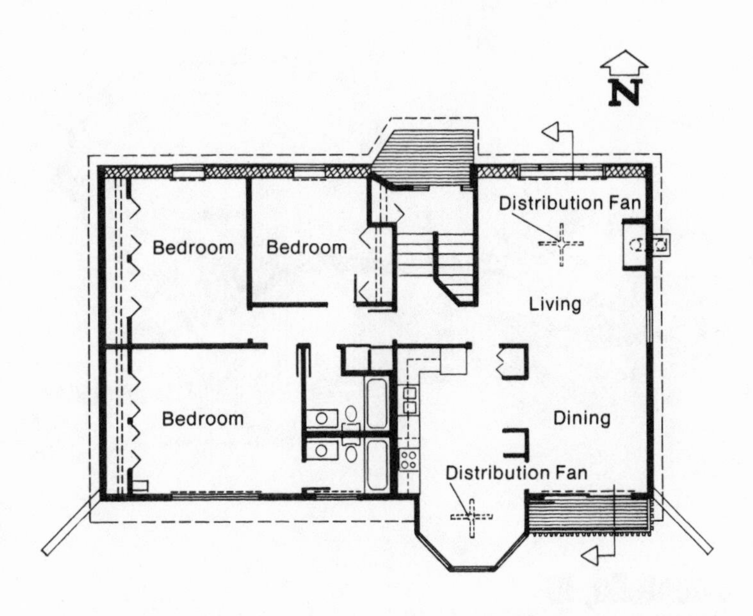
N
Distribution Fan
Bedroom
Bedroom
Living
Bedroom
Dining
Distribution Fan

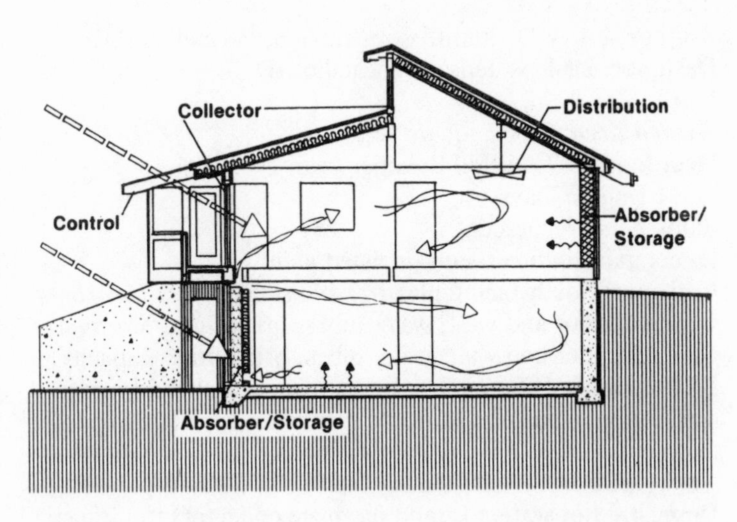
Collector
Distribution
Control
Absorber/
Storage
Absorber/Storage

Pocatello, ID

Builder: Larry D. Ratliff Construction, Pocatello, ID
Designer: Paul W. Jensen, Pocatello, ID
Price: $92,000
Heated area: 2,662 square feet
Heat load: 107 million Btus per year
6,991 Degree Days
Solar heat: 77 percent
Direct gain, indirect gain, isolated gain
Collector: South-facing glazing, 494 square feet; *Absorbers*: Concrete floor and wall, water tubes, brick floor; *Storage*: Concrete floor and wall, water tubes, brick floor (capacity: 23,942 Btus); *Distribution*: Radiation, natural and forced convection; *Controls*: Movable insulation, shade, damper, thermostat
Back-up: Electric resistance heater (75,000 Btus per hour)
Domestic hot water: Liquid flat-plate collectors (58 square feet), 120-gallon storage

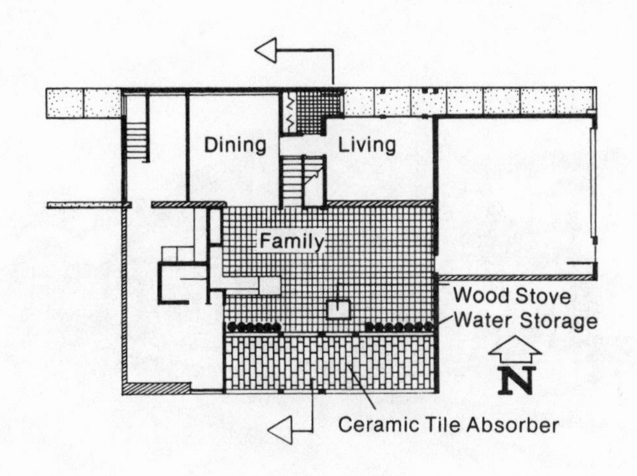
Dining
Living
Family
Wood Stove
Water Storage
N
Ceramic Tile Absorber

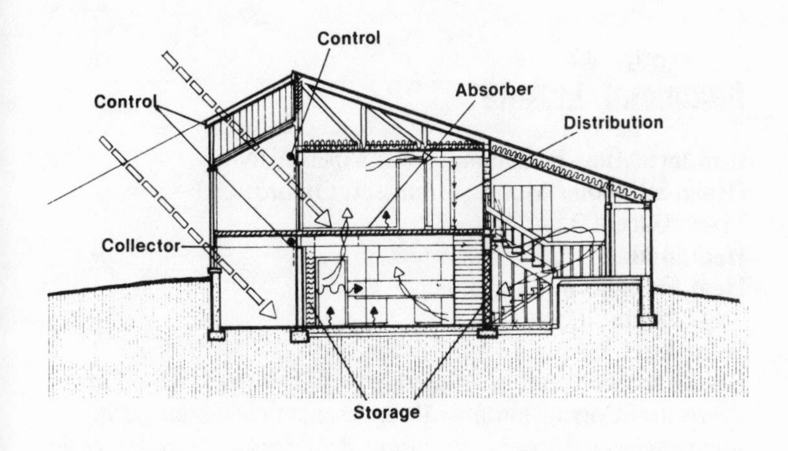
Control
Control
Absorber
Distribution
Collector
Storage

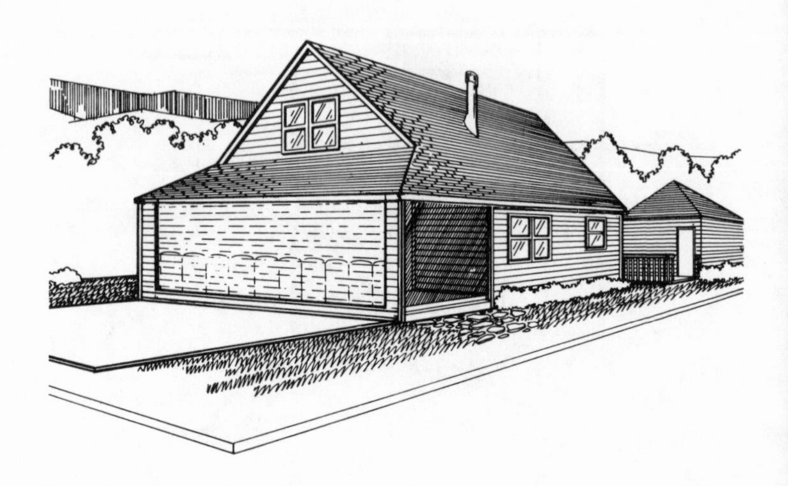

Richmond, Indiana

Builder: Miles-Richmond, Inc., Liberty, IN
Designer: Fuller Moore, Architect, Oxford, OH
Price: $40,000
Heated area: 1,005 square feet
Heat load: 40 million Btus per year
5,611 Degree Days
Solar heat: 63 percent
Isolated gain, sun tempering
Collector: Corrugated fiberglass greenhouse glazing, 256 square feet; *Absorbers*: 55-gallon steel drums; *Storage*: Water in glass tubes, water in steel drums (capacity: 5,747 Btus); *Distribution*: Radiation, natural and forced convection; *Controls*: Sliding glass panels, insulating shades, thermostat
Back-up: 51,000 Btu-per-hour wood stove; 30,600 Btu-per-hour electric resistance baseboard heaters
Cooling: Earth tubes; convection; mechanical assistance

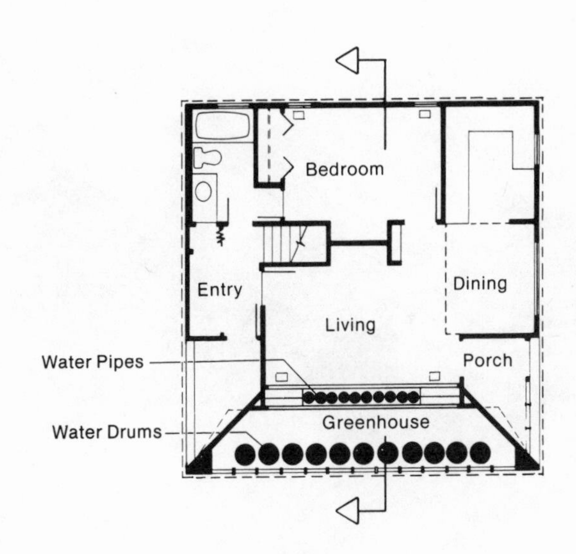
Bedroom
Entry
Living
Dining
Porch
Water Pipes
Water Drums
Greenhouse

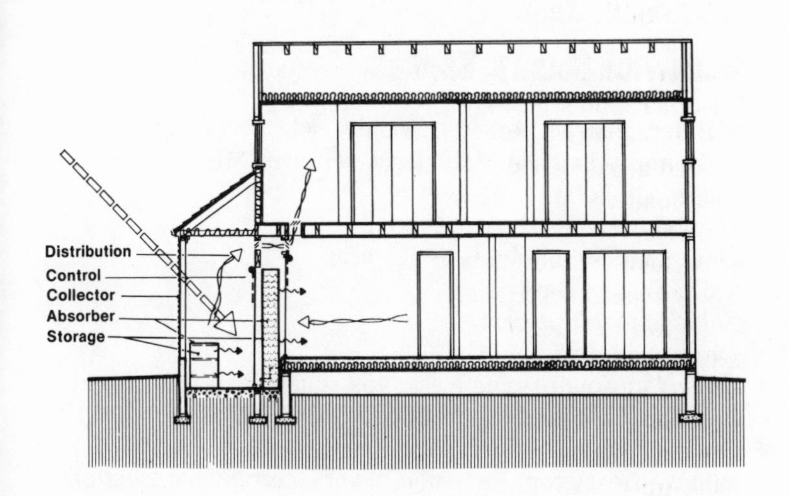
Distribution
Control
Collector
Absorber
Storage

Sedgwick County, Kansas

Builder: John M. Roberts, Wichita, KS
Designer: Allen and Mahone, Watertown, ME
Price: $132,000
Heated area: 3,411 square feet
Heat load: 99 million Btus per year
4,620 Degree Days
Solar heat: 63 percent
Direct gain, indirect gain
Collector: South-facing glass, 469 square feet; *Absorbers*: Darkened north wall surface, ceramic tile floor; *Storage*: Tile and concrete mass floor, thermal mass walls (capacity: 10,300 Btus); *Distribution*: Radiation, natural convection; *Controls*: Window Quilts, fixed overhang, removable louvers, canvas shade

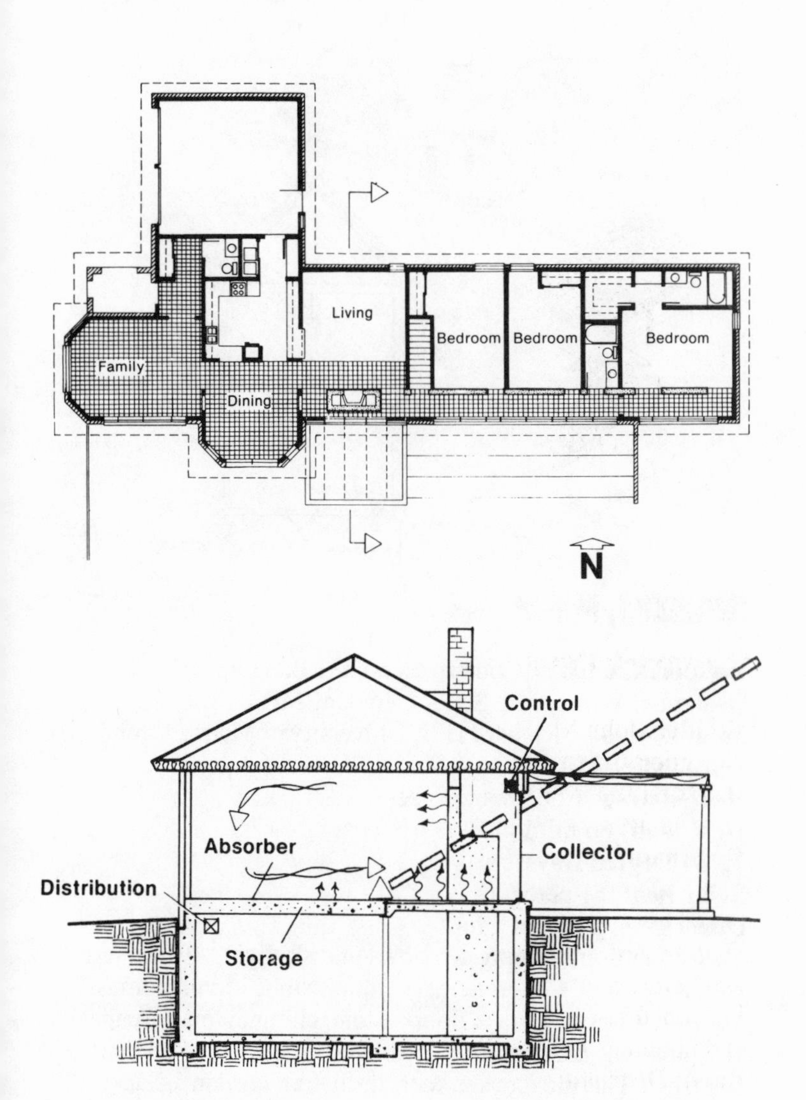
Living
Bedroom
Bedroom
Bedroom
Family
Dining
N
Control
Absorber
Collector
Distribution
Storage

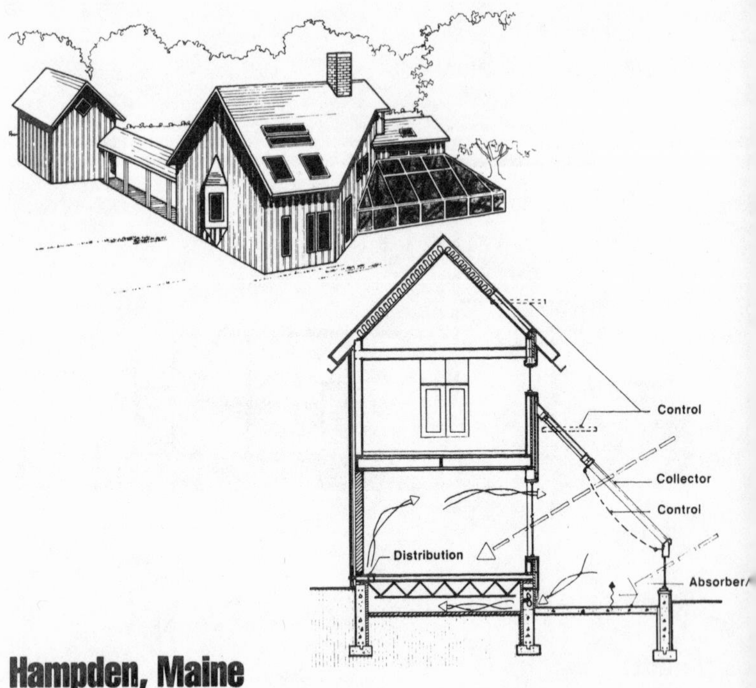

Hampden, Maine

Builder: Campbell Construction Co., Bangor, ME
Designer: William R. Sepe, Camden, ME
Solar designer: Richard Hill, University of Maine, Orono, ME
Price: $109,250
Heated area: 1,585 square feet
Heat load: 66 million Btus per year
7,784 Degree Days
Solar heat: 61 percent
Direct gain, isolated gain
Collector: South-facing windows and skylight, 498 square feet; *Absorber*: Concrete mass floors, stone chimney mass; *Storage*: Concrete mass floors, stone chimney mass (capacity: 11,314 Btus); *Distribution*: Radiation, forced convection; *Controls*: Thermostat, damper, insulating curtain, shades
Back-up: Gas wall furnaces (12,600 Btus per hour), electric baseboard heaters (12,750 Btus per hour)

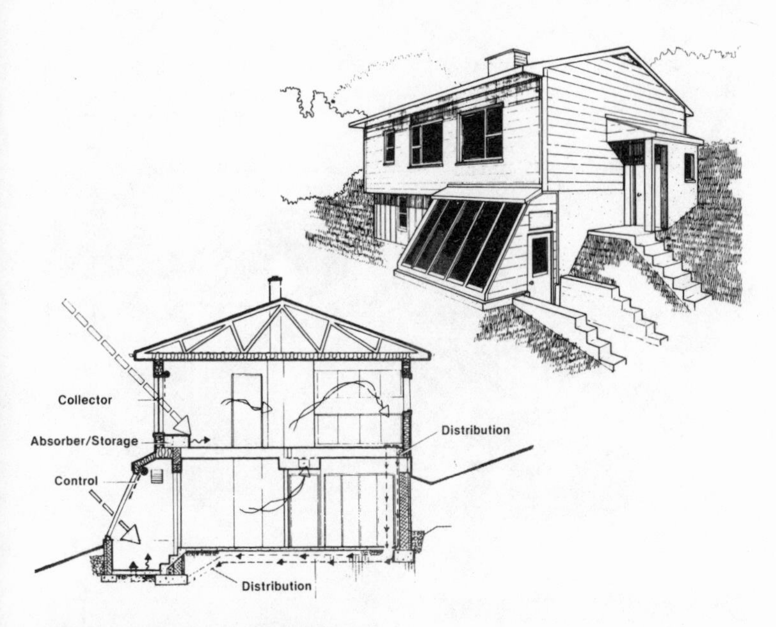

Libertytown, Maryland

Builder: Hartman, Briddell Watkins, Rockville, MD
Designer: Landon M. Proffitt, Frederick, MD
Solar designer: Solar Energy Scientific and Technical Services, Frederick, MD
Price: $75,000
Heated area: 1,675 square feet
Heat load: 81 million Btus per year
5,060 Degree Days
Solar heat: 47 percent
Direct gain, indirect gain (Trombe wall), isolated gain
Collector: South-facing glazing, mass wall glazing, greenhouse glazing, 301 square feet; *Absorber*: Concrete slab, block mass wall; *Storage*: Concrete slab, block mass wall (capacity: 4,950 Btus); *Distribution*: Radiation, convection; *Controls*: Operable vents and windows, movable insulation, sunshades
Back-up: 19,500 Btu per hour electric air-to-air heat pump, wood stove

Myersville, Maryland

Builder: M.S. Milliner Construction, Inc., Myersville, MD
Designer: Malcolm B. Wells, A.I.A., Brewster, MA
Solar designer: Solar Energy Systems and Products, Inc., Emmitsburg, MD
Price $120,000
Heated area: 2,036 square feet
Heat load: 46 million Btus per year
5,087 Degree Days
Solar heat: 63 percent
Direct gain, sun-tempering, isolated gain
Collector: South-facing clerestory windows and glazing, 248 square feet; *Absorber*: Concrete floors and wall; *Storage*: Concrete floors and wall (capacity: 18,100 Btus); *Distribution*: Radiation, natural and forced convection; *Controls*: Roll-down aluminum shades, overhangs
Back-up: Air-to-air heat pump (16,000 Btus per hour), woodburning stove, electric strip heaters
Cooling: Earth-cooled air from underground tubes, natural and forced ventilation

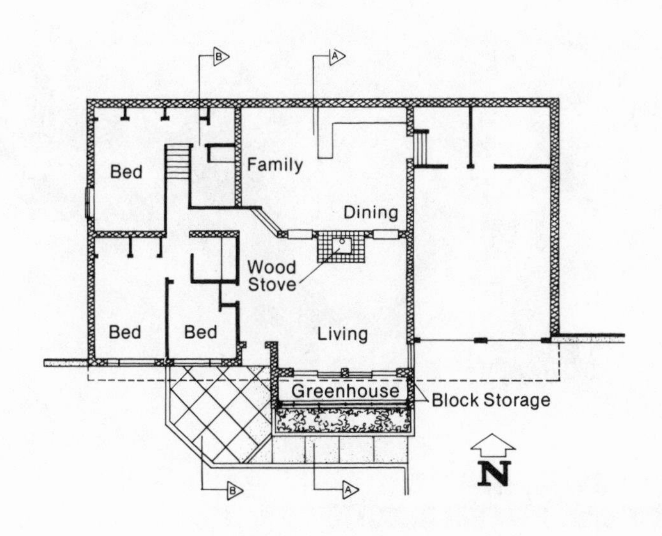
B
A
Bed
Family
Dining
Wood
Stove
Bed
Bed
Living
Greenhouse
Block Storage
N
B
A

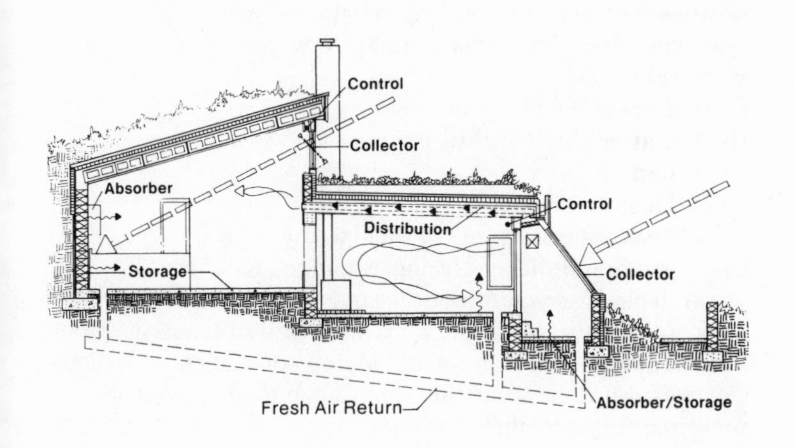
Control
Collector
Absorber
Control
Distribution
Storage
Collector
Fresh Air Return
Absorber/Storage

Foxboro, Massachusetts

Builder: Orlando Homes, Inc., Foxboro, MA
Designer: The Ehrenkrantz Group, New York, NY
Price: $130,000
Heated area: 2,443 square feet
Heat load: 80 million Btus per year
4,788 Degree Days
Solar heat: 46 percent
Direct gain, isolated gain, sun-tempering
Collector: South-facing glazing, sliding glass doors, 331 square feet; *Absorber*: Floor tiles over concrete slab floor, brick wall; *Storage*: Concrete floor, brick wall (capacity 8,009 Btus); *Distribution*: Natural and forced convection, radiation; *Controls*: Movable insulation (pull-down shade), overhangs, sun-screens, vents, thermostat
Back-up: Electric resistance heater, air-to-water heat pump (44,000 Btus per hour)
Domestic hot water: 36 square feet liquid flat-plate collectors, 100-gallon storage

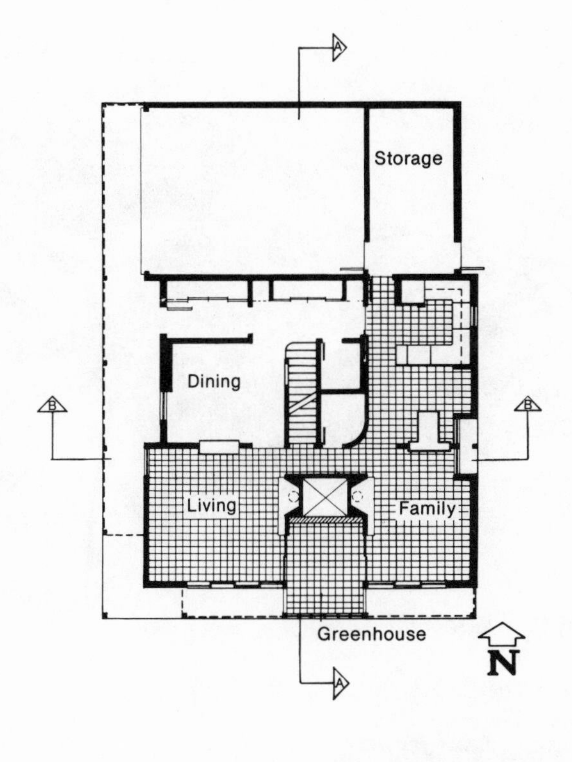
A
Storage
Dining
B
B
Living
Family
Greenhouse
N
A

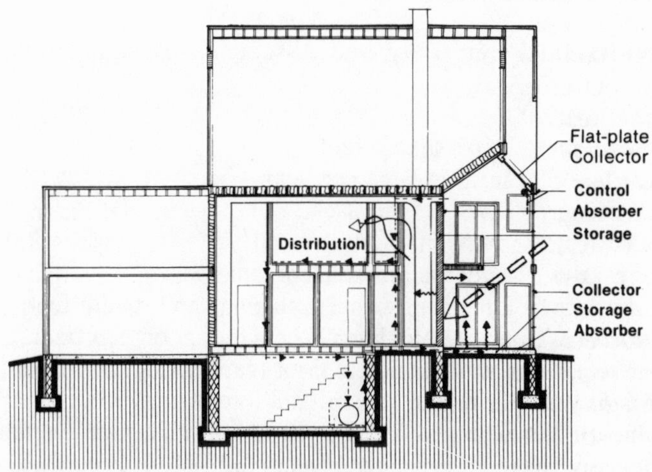
Flat-plate Collector
Control
Absorber
Storage
Distribution
Collector
Storage
Absorber

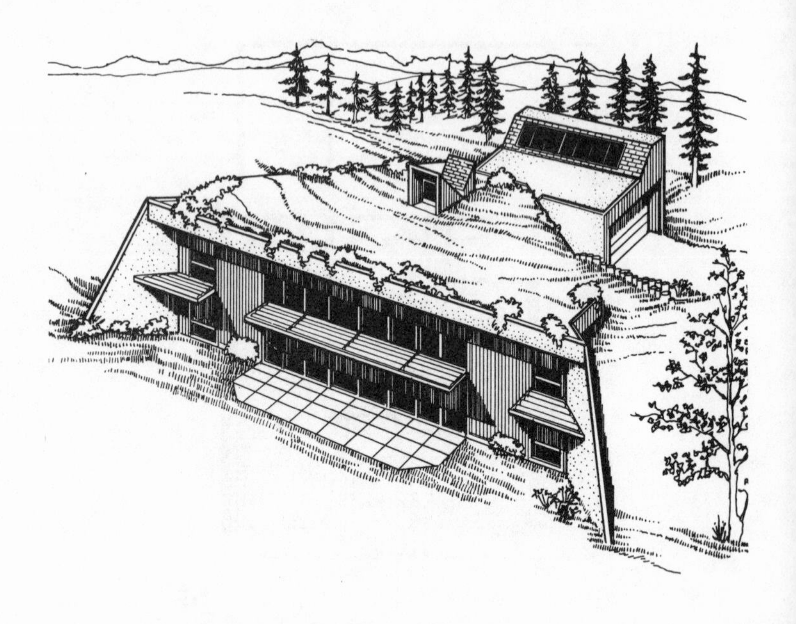

Corcoran, Minnesota

Builder/Designer: Berg and Associates, Design/Builders, Plymouth, MN
Price: $120,000
Heated area: 1,665 square feet
Heat load: 77 million Btus per year
8,054 Degree Days
Solar heat: 83 percent
Direct gain, indirect gain, isolated gain
Collectors: South-facing panels, glazing, 560 square feet; *Absorbers*: Concrete block wall, concrete floor; *Storage*: Concrete block wall, concrete floor (capacity: 45,116 Btus); *Distribution*: Radiation, natural and forced convection; *Controls*: Movable insulation on Trombe walls, roof overhang
Back-up: Electric resistance heaters (30,000 Btus per hour)

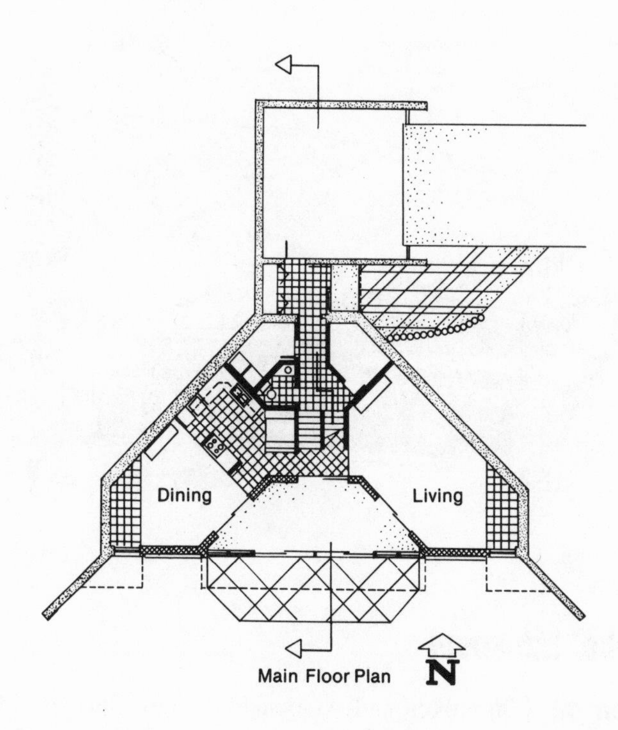

Main Floor Plan

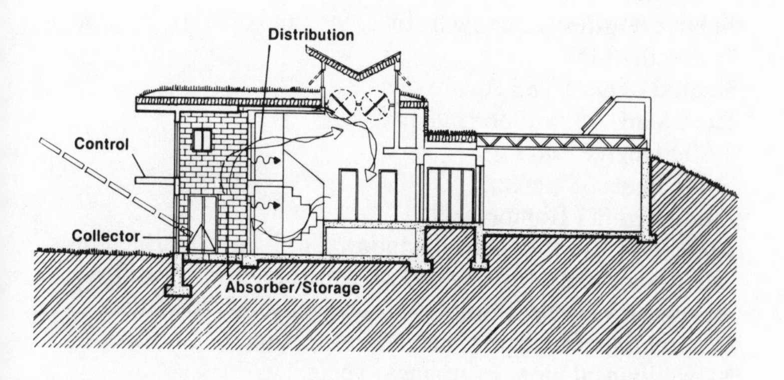

Eureka, Missouri

Builder: G.T. Kinniken and Associates, House Springs, MO
Designer: Robert Lutz Architects, St. Louis, MO
Solar designer: Ener-Tech, Inc., St. Louis, MO
Price: $115,000
Heated area: 1,783 square feet
Heat load: 48 million Btus per year
4,900 Degree Days
Solar heat: 58 percent
Indirect gain (Trombe wall)
Collectors: Double-glazed windows, reflective marble chips, 292 square feet; *Absorber*: Dark concrete Trombe wall; *Storage*: Concrete Trombe wall (capacity: 54,264 Btus); *Distribution*: Radiation, natural convection; *Controls*: Movable insulation, overhangs, vents
Back-up: Electric resistance heating, fireplace
Cooling: Earth sheltering, convection

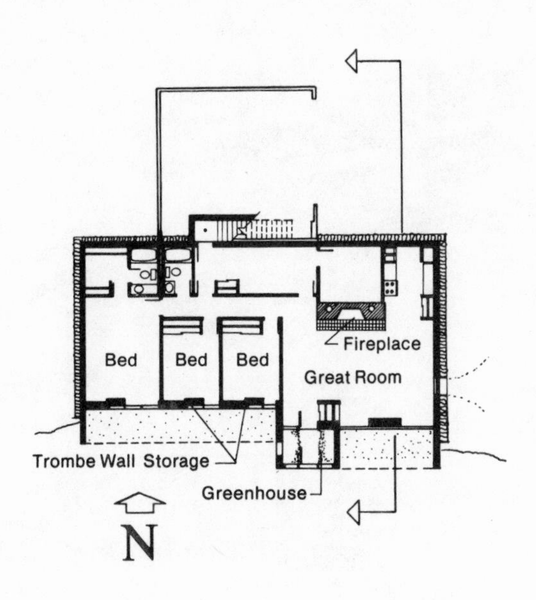
Fireplace
Bed
Bed
Bed
Great Room
Trombe Wall Storage
Greenhouse
N

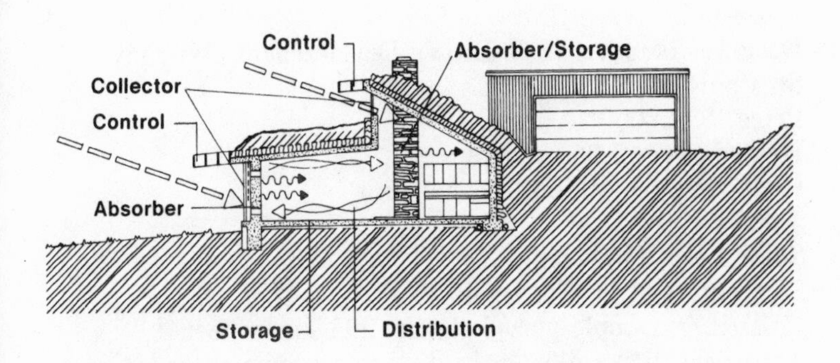
Control
Absorber/Storage
Collector
Control
Absorber
Storage
Distribution

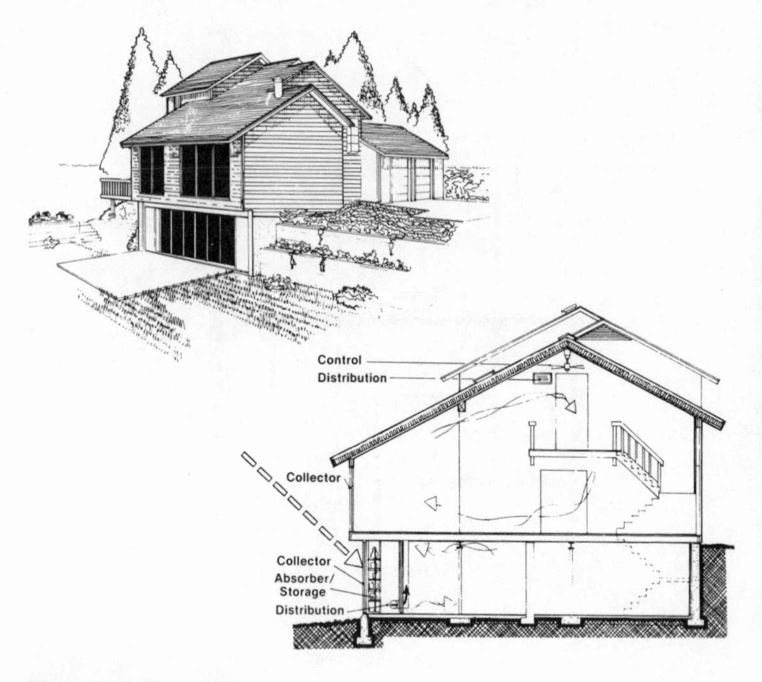

New Melle, MO

Designer/Builder: N.O. Brown Development Company, St. Louis, MO
Price: $89,000
Heated area: 2,050 square feet
Heat load: 63 million Btus per year
4,900 Degree Days
Solar heat: 53 percent
Sun tempering, indirect gain
Collectors: Double-glazed panels, 293 square feet; *Absorbers*: Water-filled polyethylene containers; *Storage*: Water-filled polyethylene containers (capacity: 4,725 Btus); *Distribution*: Radiation, natural and forced convection; *Controls*: Return air registers, roll-down shades, overhangs
Back-up: 51,000-Btu heat pump, wood stove
Domestic hot water: Optional pre-heat

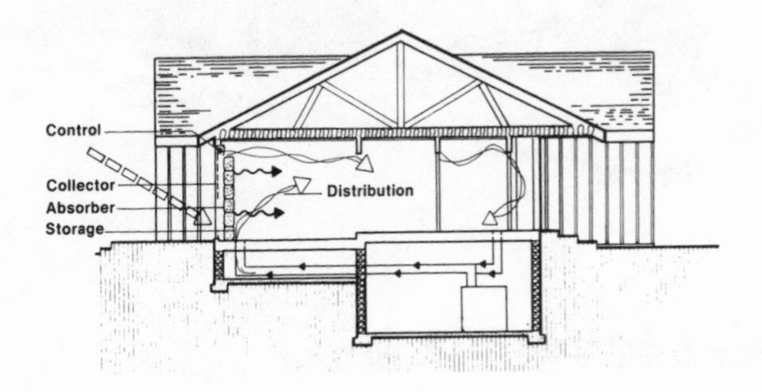

Blackwood, New Jersey

Builder: Diamond Crest, Inc., Blackwood, NJ
Designer: Princeton Energy Group, Princeton, NJ
Heated area: 1,691 square feet
Heat load: 42 million Btus per year
4,812 Degree Days
Solar heat: 72 percent
Indirect gain, sun-tempering
Collector: Double-glazed windows, acrylic panels, 296 square feet; *Absorber*: Water wall surface; *Storage*: Water wall (capacity: 8,075 Btus); *Distribution*: Radiation, forced convection; *Controls*: Movable insulation, awnings
Back-up: Gas furnace (50,000 Btus per hour)
Domestic hot water: Liquid flat-plate collectors (84 square feet), 80-gallon storage

Manzano Springs, New Mexico

Builder: Enecon, Inc., Santa Fe, NM
Designer: Clark-Germanas Architects, Santa Fe, NM
Solar designer: Barkmann Engineering, Santa Fe, NM
Price range: $75,000 to $80,000
Heated area: 1,500 square feet
Heat load: 52 million Btus per year
5,780 Degree Days
Solar heat: 75 percent
Direct gain, indirect gain
Collectors: Double-glazed windows, double-glazed clerestory windows, 255 square feet; *Absorbers*: Water tubes, brick floor over concrete slab; *Storage*: Water tubes, concrete slab (capacity: 13,917 Btus); *Distribution*: Radiation, natural convection; *Controls*: Insulated venetian blinds, shutters and curtains, fixed overhangs, earth berms
Back-up: Woodburning stove, electric resistance heat
Domestic hot water: Active solar collector
Cooling: Natural ventilation

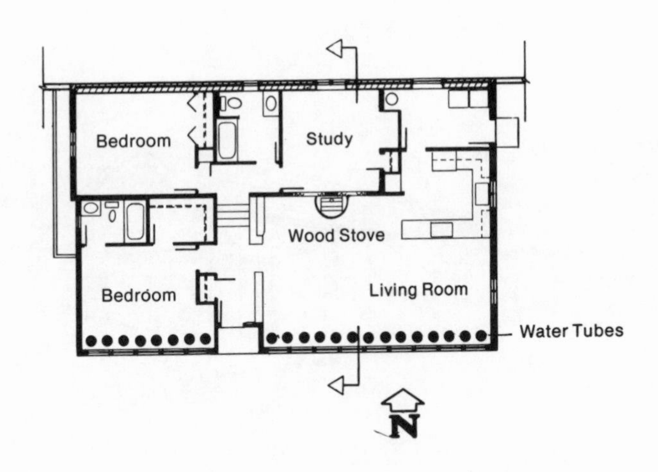
Bedroom
Study
Wood Stove
Bedroom
Living Room
Water Tubes
N

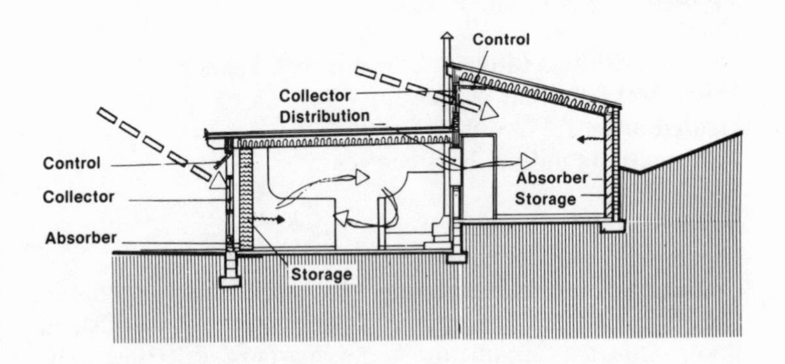
Control
Collector
Distribution
Control
Collector
Absorber
Storage
Absorber
Storage

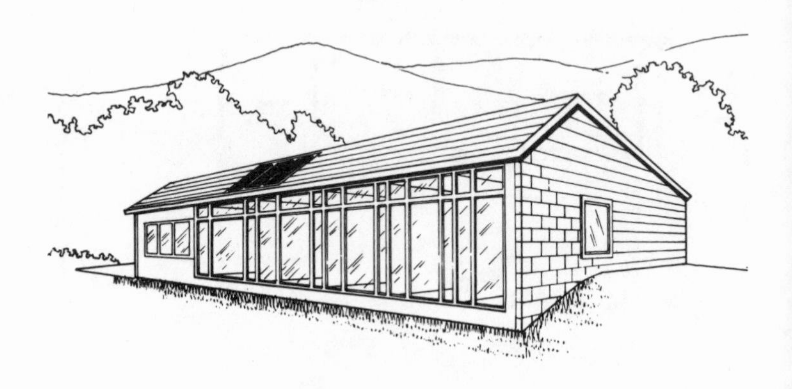

Putnam Valley, New York

Designer/Builder: Robert Brown Butler, Katonah, NY
Price: $60,000 (plus land)
Heated area: 1,473 square feet
Heat load: 42 million Btus per year
5,732 Degree Days
Solar heat: 90 percent
Direct gain, indirect gain
Collector: South-facing glazing, clerestory windows, 368 square feet; *Absorber*:Ceramic tile floors, masonry walls and floors; *Storage*: Ceramic tile floors, masonry walls and floor (capacity: 24,583 Btus); *Distribution*: Natural and forced convection; *Controls*: Thermal shutters that operate like garage doors, overhang, thermostats
Back-up: Woodburning stove, electric resistance heaters
Domestic hot water: Three-panel active system (84 square feet)

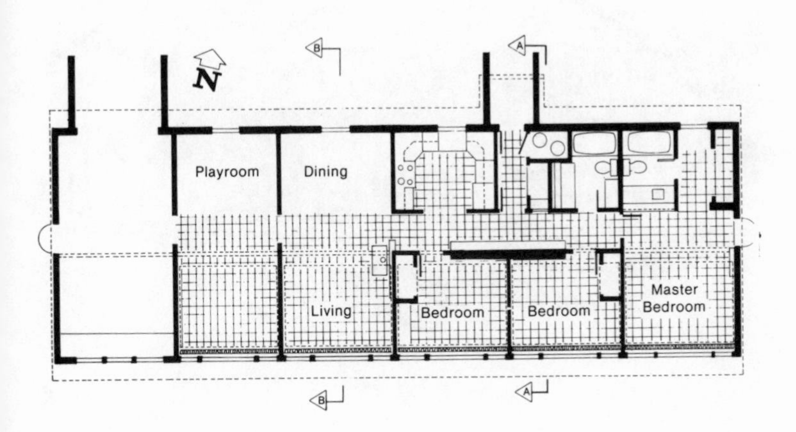
N
B
A
Playroom
Dining
Living
Bedroom
Bedroom
Master Bedroom

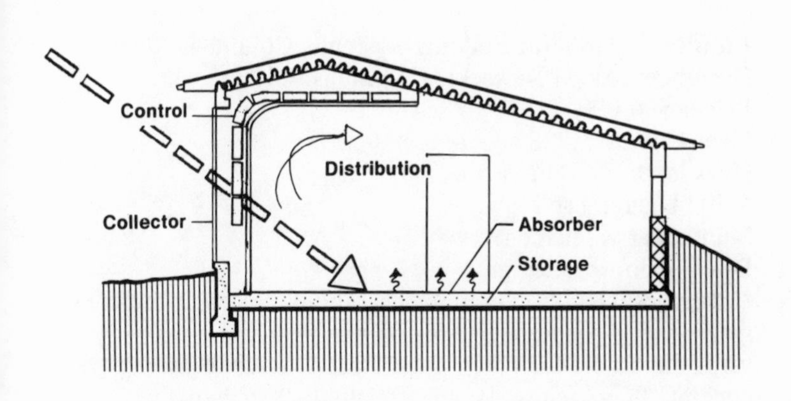
Control
Distribution
Collector
Absorber
Storage

Dublin, Ohio

Builder: Solartherm Building Systems, Columbus, OH
Designer: Joseph Kawecki, Columbus, OH
Price: $90,000
Heated area: 1,828 square feet
Heat load: 65 million Btus per year
5,702 Degree Days
Solar heat: 41 percent
Direct gain, isolated gain
Collectors: Double-glazed panels, sliding doors, greenhouse glazing, 524 square feet; *Absorbers*: Greenhouse masonry wall, tile-covered concrete floor slab; *Storage*: Masonry wall, concrete floors (capacity: 13,416 Btus); *Distribution*: Radiation, natural and forced convection; *Controls*: Thermostats, insulating blinds and shutters, overhangs
Back-up: Electric resistance heater (20,000 Btus per hour); woodburning stove

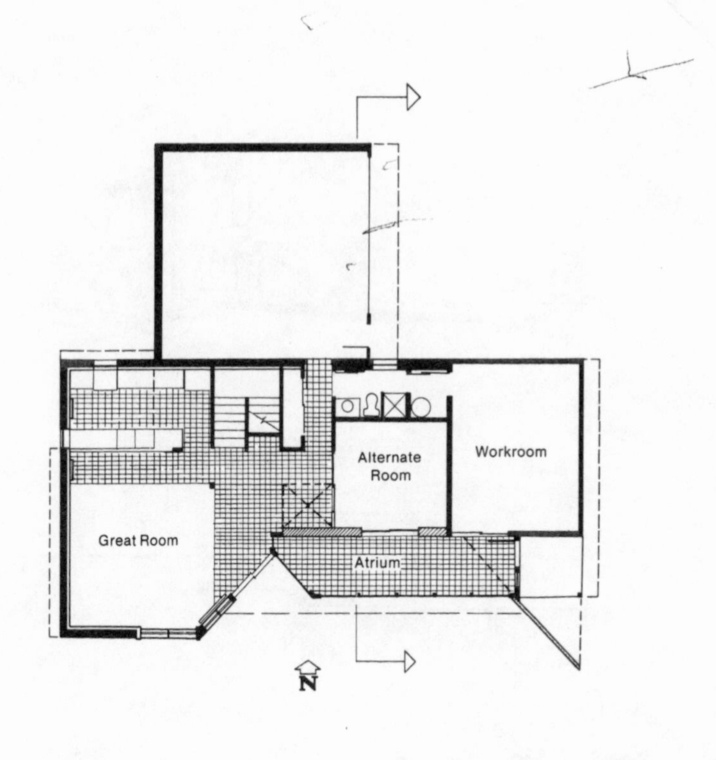
Alternate Room
Workroom
Great Room
Atrium
N

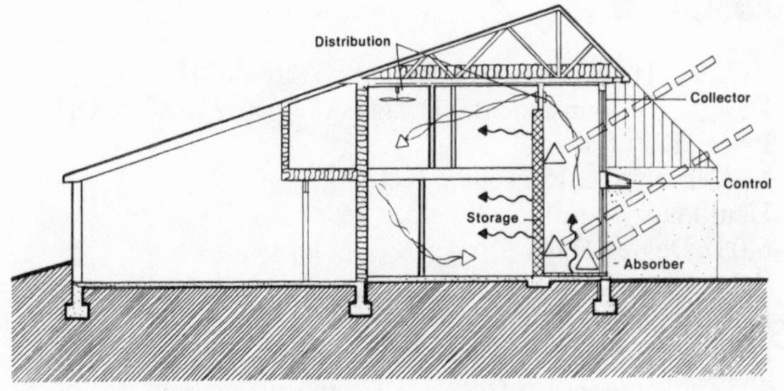
Distribution
Collector
Control
Storage
Absorber

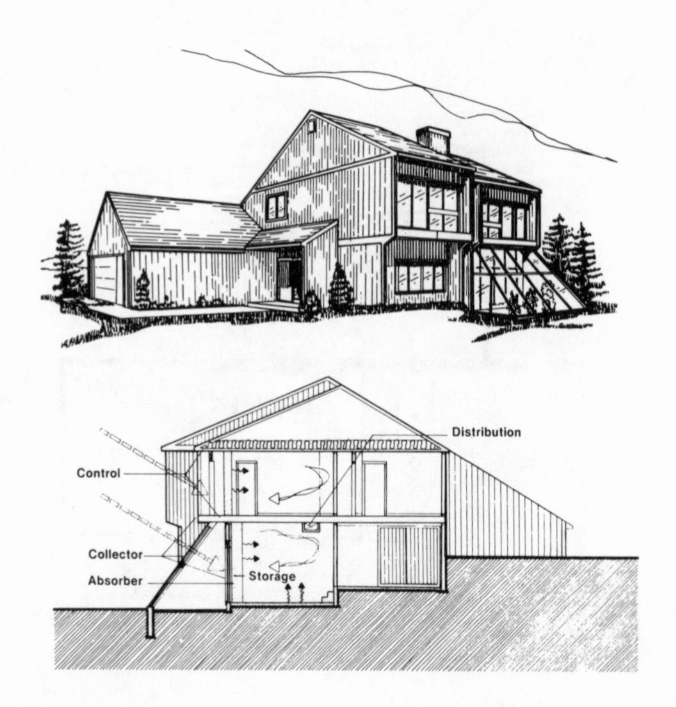

Kent, Ohio

Builder: Huth Westwood Builders, Akron, OH
Designer: Environmental Design Alternatives, Kent, OH
Price: $130,000
Heated area: 2,158 square feet
Heat load: 76 million Btus per year
6,037 Degree Days
Solar heat: 46 percent
Isolated gain, sun tempering
Collectors: Single and double-glazed windows and doors, 423 square feet; *Absorbers*: Water tubes, ceramic tile floor; *Storage*: Water tubes and mass floors (capacity: 16,345 Btus); *Distribution*: Radiation, natural and forced convection; *Controls*: Insulating shutters, fixed and retractable overhangs
Back-up: Woodburning stove, gas heat

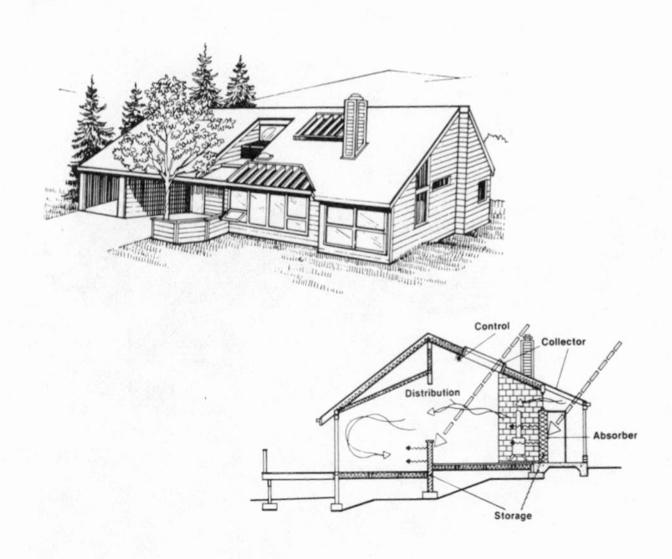

Springfield, Oregon

Builder: Paz Peterson McRae, Inc., Eugene, OR
Designer: Artemio Paz, Jr., Springfield, OR
Price: $100,000
Heated area: 1,719 square feet
Heat load: 65 million Btus per year
4,739 Degree Days
Solar heat: 60 percent
Direct gain, indirect gain, isolated gain
Collectors: Skylight, Trombe wall glazing, greenhouse glazing, 320 square feet; *Absorbers*: Concrete block wall, concrete block Trombe wall, greenhouse concrete slab floor; *Storage*: Concrete block wall, concrete block Trombe wall, greenhouse greenhouse slab floor (capacity: 14,612 Btus); *Distribution*: Radiation, natural convection; *Controls*: Movable shading devices
Back-up: Electric resistance wall units; wood stove
Domestic hot water: "Breadbox" collector, 60-gallon tanks
Cooling: Natural cross-ventilation, convection

Winston, Oregon

Builder: James L. Richey, Jr., Winston, OR
Designer: Sunwood Building and Design, Roseburg, OR
Price: $75,000
Heated area: 1,652 square feet
Heat load: 71 million Btus per year
4,496 Degree Days
Solar heat: 30 percent
Direct gain
Collectors: Double-glazed windows, 246 square feet; *Absorbers*: Concrete floors, mass walls, brick tile floor; *Storage*: Concrete floor, concrete block walls (capacity: 17,183 Btus); *Distribution*: Radiation, natural and forced convection; *Controls*: Window shutters, thermostat, overhang, movable insulation
Back-up: Electric furnace, wood stove
Domestic hot water: Active closed loop solar system

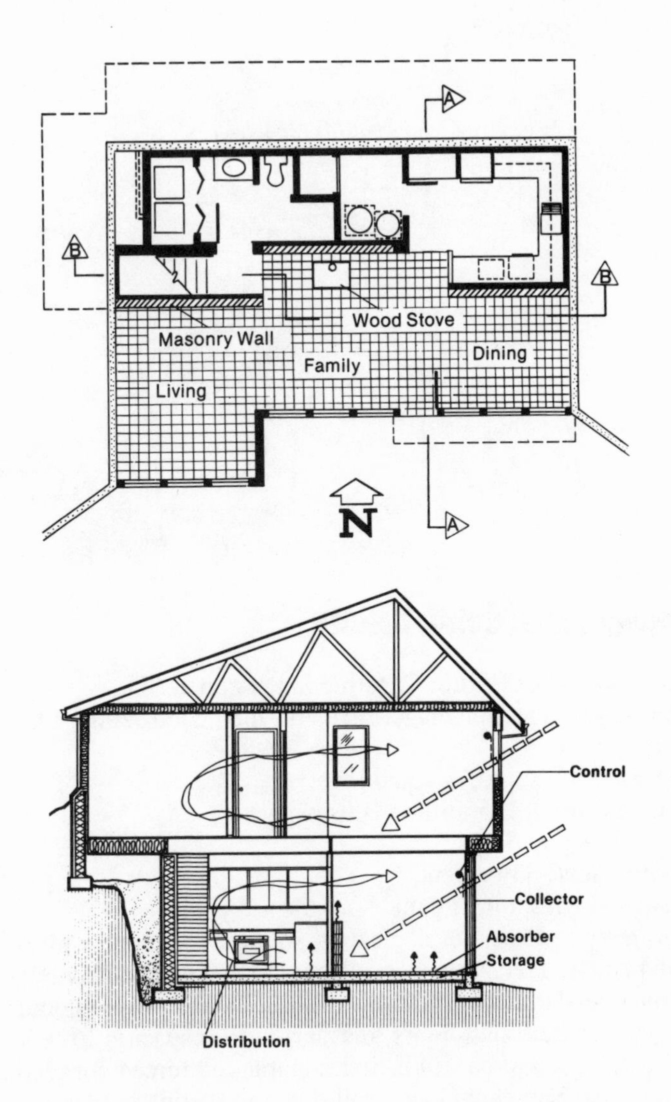
A
B
B
Wood Stove
Masonry Wall
Family
Dining
Living
N
A
Control
Collector
Absorber
Storage
Distribution

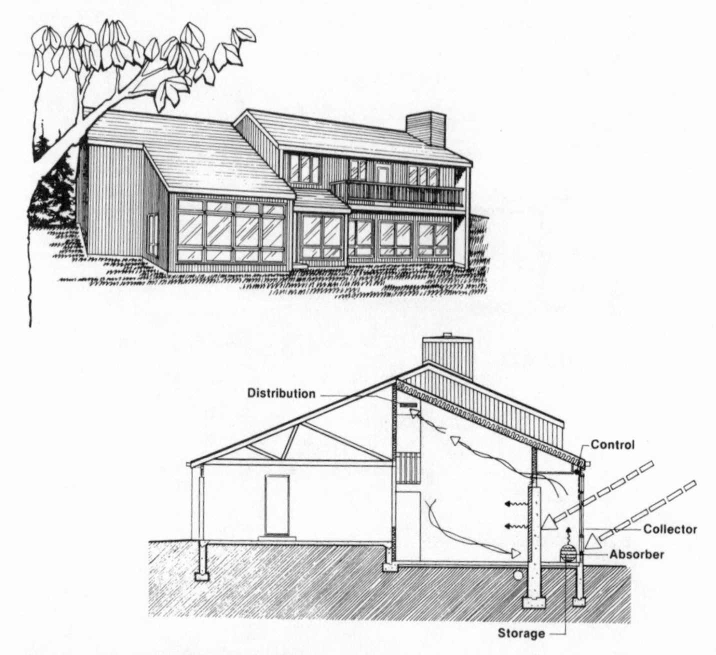

Sioux Falls, South Dakota

Builder: Dick Sorum Construction, Renner, SD
Designer: The Spitznagel Partners, Inc., Sioux Falls, SD
Price $100,000
Heated area:: 2,074 square feet
Heated load: 125 million Btus per year
7,839 Degree Days
Solar heat: 33 percent
Isolated gain, direct gain
Collectors: Greenhouse glazing, south-facing double glazing, 364 square feet; *Absorbers*: Water storage tank surface, concrete slab floors, masonry walls; *Storage*: Water storage tank, concrete slab floors, masonry wall (capacity: 16,028 Btus); *Distribution*: Radiation, natural and forced convection; *Controls*: Vents, dampers, roll-down insulating curtain and shades, greenhouse overhang
Back-up: Gas forced-air heating system

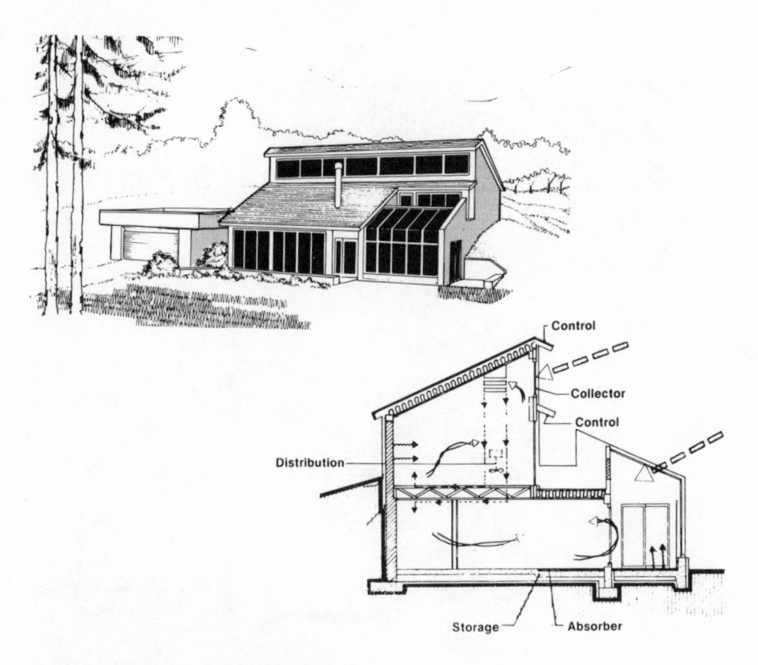

Oak Ridge, Tennessee

Builder: Winston Cox Admiral Construction Co., Oak Ridge, TN
Designer: Ron E. Barstow, Knoxville, TN
Price: $85,000
Heated area: 1,939 square feet
Heat load: 54 million Btus per year
3,949 Degree Days
Solar heat: 83 percent
Direct gain, sun tempering
Collectors: Double-glazed windows, 548 square feet; *Absorber*: Concrete brick floor; *Storage*: Concrete brick floor, cement block wall (capacity: 65,474 Btus); *Distribution*: Radiation, natural and forced convection; *Controls*: Insulating shutters, shades, overhangs
Back-up: Air-to-air heat pump (18,000 Btus per hour); wood stove

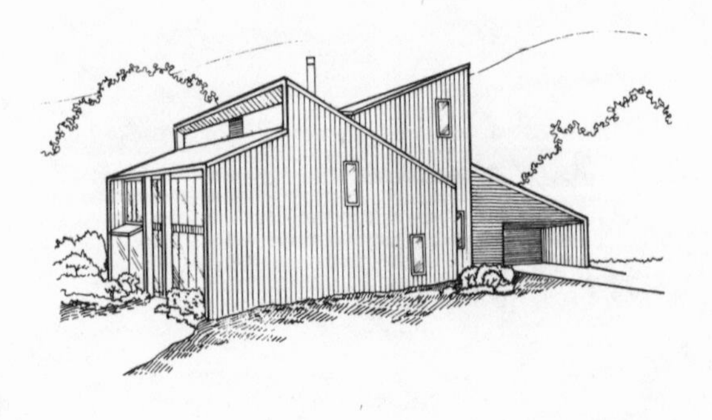

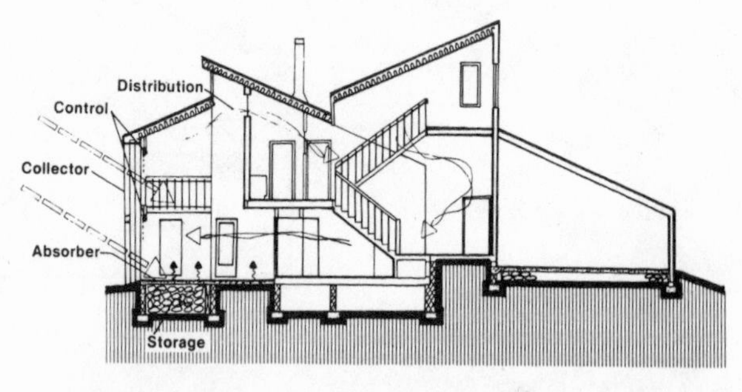

Memphis, Tennessee

Builder: R.N. Stanley Construction Co., Memphis, TN
Designer: Solar Energy & Environmentally Designed Systems, Memphis, TN
Price: $90,000
Heated area: 2,560 square feet
Heat load: 72 million Btus per year
3,232 Degree Days
Solar heat: 78 percent
Direct gain, sun tempering
Collectors: Double-glazed windows, 655 square feet; *Absorber*: Concrete and stone floor; *Storage*: Concrete and stone floor (capacity: 21,388 Btus); *Distribution*: Radiation, natural and forced convection; *Controls*: Overhangs, shades, fans, curtains
Back-up: Gas forced-air furnace (48,000 Btus per hour)

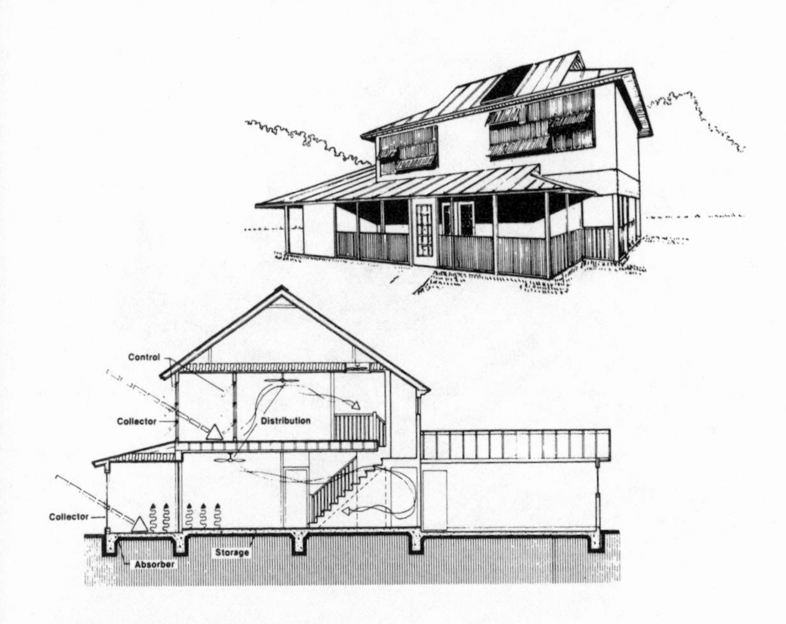

College Station, Texas

Builder: Dwayne Rhea Construction, Inc., College Station, TX
Designer: Entek Associates, Inc., College Station, TX
Solar designer: Paul H. Woods
Price: $76,000
Heated area: 1,744 square feet
Heat load: 36 million Btus per year
1,617 Degree Days
Solar heat: 50 percent
Isolated gain
Collector: South-facing greenhouse glazing, 467 square feet; *Absorbers and storage*: Concrete floor and wall; *Distribution*: Radiation, natural and forced convection; *Controls*: Movable insulating panels, vents, roof overhangs, aluminum shade screens
Back-up: Electric heater (24,000 Btus per hour)
Domestic hot water: Liquid flat-plate collectors (38 square feet), 80-gallon storage

Port Townsend, Washington

Designer/builder: Burnham Construction, Port Townsend, WA
Price: $92,000
Heated area: 1,596 square feet
Heat load: 63 million Btus per year
5,330 Degree Days
Solar heat: 65 percent
Direct gain, indirect gain (Trombe wall), isolated gain
Collectors: Trombe wall glazing, greenhouse, skylight, glass doors, 278 square feet; *Absorbers*: Surface of Trombe wall, mass walls and floors; *Storage*: Trombe wall, greenhouse wall, living room wall, concrete slab floors (capacity: 9,885 Btus); *Distribution*: Radiation, natural and forced convection; *Controls*: Fixed overhangs, movable shade, louvers on greenhouse, earth berms
Back-up: Woodburning stove, electric baseboard heaters
Domestic hot water: 55 square feet of active flat plate collectors

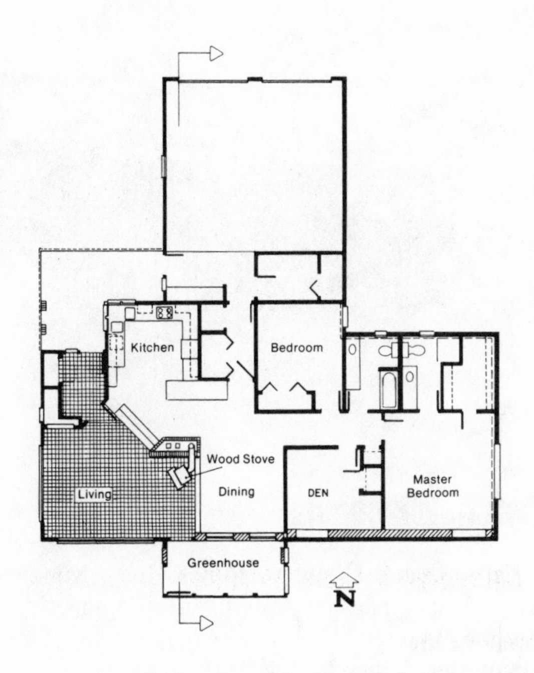
Kitchen
Bedroom
Wood Stove
Living
Dining
DEN
Master
Bedroom
Greenhouse
N

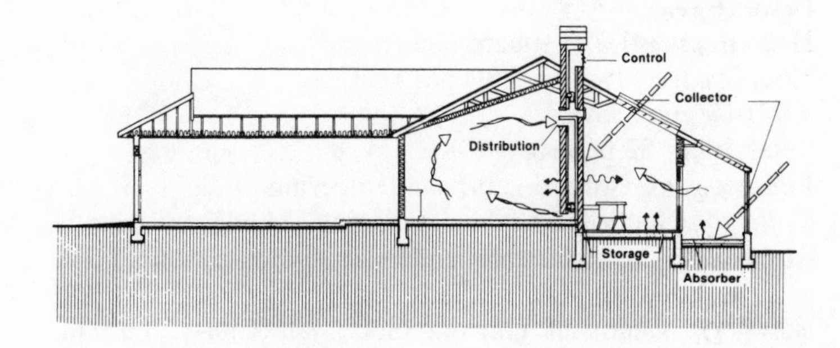
Control
Collector
Distribution
Storage
Absorber

Middleton, Wisconsin

Builder: Northland Country Homes, Inc., Middletown, WI
Designer: North Design Architecture/Engineering/Planning, Middletown, WI
Solar designer: Bruce D. Kieffer
Price: $79,500
Heated area: 1,788 square feet
Heat load: 84 million Btus per year
7,721 Degree Days
Solar heat: 52 percent
Indirect gain, isolated gain, sun tempering
Collector: South-facing double glazing, 343 square feet; *Absorbers*: Greenhouse concrete slab floor, water tube surface, concrete block wall; *Storage*: Greenhouse concrete slab floor, water wall, concrete block wall (capacity: 9,600 Btus); *Distribution*: Radiation, natural convection; *Controls*: Vents, registers, operable thermal doors, fixed overhangs, insulated shades
Back-up: Gas furnace (64,000 Btus per hour)
Cooling: Earth-cooled underground pipes, natural ventilation

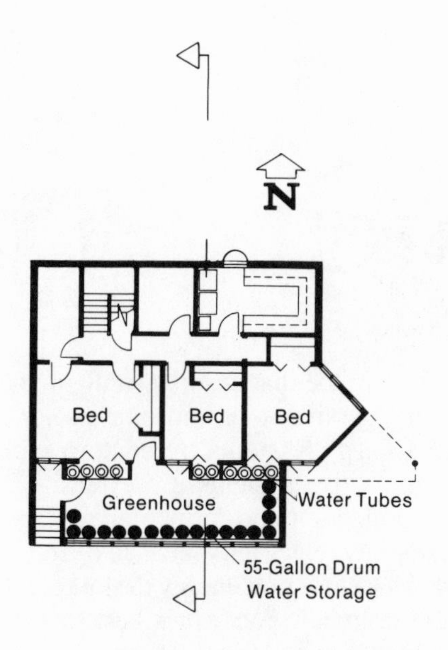
N
Bed
Bed
Bed
Greenhouse
Water Tubes
55-Gallon Drum
Water Storage

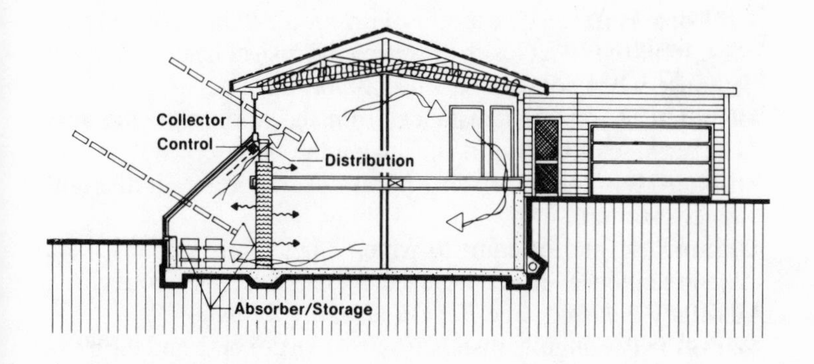
Collector
Control
Distribution
Absorber/Storage

Glossary

Absorber: The surface that absorbs sunlight and converts it to heat energy, in a passive solar house or a solar collector.
Active Solar Energy Systems: In contrast with passive solar energy systems, active systems use extensive "hardware"—i.e. flat plate collectors and large storage tanks—as well as conventional energy (electricity, etc.) to operate the system and to transfer the collected solar energy to storage, and to distribute it throughout the living space. It is best to avoid using active systems for space heating, but they are appropriate for heating household water.
Air Lock Entry: A vestibule with an air-tight door on both sides. It reduces heat loss by limiting the rush of heated air out of the house when someone is entering or leaving.
Altitude: The angular distance from the horizon to the sun. "The altitude of the sun is 45° above the horizon."
Ambient Temperature: This phrase usually refers to the outside temperature.
Atrium: An interior court to which other rooms open. Like a solar greenhouse, it can be used for passive solar collection.
Auxiliary Heating: See *Back-up Energy System*.
Azimuth: The angular distance between true south and the point on the horizon directly below the sun.
Back-up Energy System: The system that supplements the solar system when necessary. A back-up system using electricity, oil, gas, or wood should be capable of meeting the demand for heat during any period when the solar heating system cannot.

Btu: British thermal unit, a basic heat measurement equivalent to the amount of heat needed to raise the temperature of 1 pound of water by 1° Fahrenheit.
Building Envelope: The parts of a building (walls, roof, floors) that enclose heated spaces.
Charge: Putting heat into storage through the absorption of radiant energy or convective heat transfer.
Clerestory: Pronounced CLEAR STORY, a window placed high in a wall near the eaves, used for light, heat gain, and ventilation.
Collection: The act of trapping solar radiation and converting it to heat.
Collector Aperture: The glazed opening that admits solar radiation.
Comfort Zone: The range of temperature and humidity within which most people feel comfortable.
Conduction: The flow of heat between a hotter material and a colder material when the two materials are in direct physical contact.
Controls: The assemblage of devices used to regulate the processes of collecting, transporting, storing, and using solar energy.
Convection (forced): Commonly, the transfer of heat by means of the forced flow of air or water.
Convection (natural): The movement of heat that occurs according to natural laws, whereby a warm fluid rises and a cool fluid sinks under the influence of gravity.
Cooling Pond: A large body of water that loses heat from its surface, largely by evaporation, but also by convection and radiation.
Damper: A control that permits, prevents, or controls the passage of air through a duct.
Deciduous Trees: Trees that shed their leaves each fall.
Degree Days (DD): A measure of the need for heat, based on the assumption that temperature should be maintained at 65°F for comfort. Generally, degree days accumulate in the following way: If the average outdoor temperature during a given day is 50°F, fifteen degrees lower than 65°F, then the number of degree

days that accumulates on that day is fifteen (65 minus 50). Over the course of a winter, the number of degree days is added to use as a measure of the severity of winter in a particular area, city, or town.

Design Heating Load: The total heat loss from a building under the most severe winter conditions likely to occur.

Diffuse Radiation: Indirect, scattered sunlight that casts no shadow.

Direct Radiation: Sunlight that casts shadows, also called beam radiation.

Direct Solar Gain: The most common type of passive solar heating, direct solar gain occurs when sunlight is allowed to enter the living space and collected there as heat. The solar radiation usually enters a room through large areas of south-facing glass. The heat is absorbed and stored by masonry floors and walls, or by other materials placed so that they are warmed by the sun. One advantage of direct gain is that it allows immediate warming of the space when the sun is shining, as well as providing natural lighting and an outdoor view. It has the potential, on the other hand, to overheat the living space occasionally.

Discharge: Removing heat from storage by radiation or convective heat transfer.

Distribution: The movement of collected heat, from collectors or storage, throughout living areas.

Diurnal Temperature Range: The variation in outdoor temperature between day and night.

Double-glazed: A double layer of glass or plastic window or collector glazing material.

Earth Berm: A mound of dirt pushed against a building wall to stabilize interior temperatures or to deflect wind.

Emittance: A measure of the tendency of a material to radiate energy to its surroundings.

Energy Audit: An inspection and account of the energy used, and the ways in which it was used, by a particular house during a particular period of time (a month, or a heating season).

Energy Conservation Features: Insulation, storm windows, weatherstripping, caulking, and similar measures that reduce

the loss of heat in winter, or its gain in summer.
Evaporative Cooling: A method of cooling that uses the evaporation of water to reduce the temperature of spaces.
Evergreen, or Coniferous Trees: Trees that do not shed leaves in the fall, and can therefore be used as windbreaks on the north side of a building site.
Fan Coil: A unit consisting of a fan and a heat exchanger that transfers heat from liquid to air, or from air to liquid. Usually it is located in a duct.
Forced-Air Heat: A conventional heat distribution system that uses a blower to circulate heated air.
Glazed Area (or Glazing): For solar collection, glazing refers to all materials that are transparent or translucent to short-wave solar radiation (light). These include glass, acrylics and other kinds of plastic, some forms of rigid fiberglass, and the like.
Heat Capacity: The ability of a material to absorb heat.
Heat Exchanger: A device that transfers heat from one fluid to another. For the purposes of heat movement, air and water are both considered to be fluids.
Heat Gain: The amount of heat that a space gains from all sources (including people, lights, machines, sunshine).
Heat Pump: A machine, powered by electricity, that can be used to heat and cool a house, based on the same principal as refrigerators and air conditioners.
Heat Storage: The preserving of heat *as* heat (not as a source of energy—like gas or oil—that can be used to produce heat later; a device or material that absorbs collected solar heat and stores it for use during periods when the sun is not shining.
Heat Storage Capacity: The amount of heat that can be stored by a given quantity of a given material.
Heat Transfer: The means by which heat moves from one material to another—conduction, convection, or radiation, or a combination of these.
Heating Load: The amount of heat required to keep a house comfortable, usually measured in Btus per hour, day, or year.
Heating Season: In the northern hemisphere, the period from early fall to late spring during which heat is needed to keep a house comfortable for its occupants.

Hybrid System: A hybrid system is one incorporating a major passive component with an active component, in which at least one of the significant thermal energy flows is by natural means, and at least one is by forced means (fans, ducts, blowers).

Indirect Solar Gain: A method of passive solar heating whereby sunlight is collected through south-facing glazing but is immediately intercepted by an absorber and storage element that separates the glazing from the living space, blocking the entrance of sunlight into the living space. The Trombe wall and the water wall are two of the most frequently used indirect gain systems.

Infiltration: The movement of outdoor air into a building through leaks, cracks, openings around windows and doors.

Insolation: The amount of solar radiation (direct, diffuse, or reflected) that strikes a surface exposed to the sky, measured in Btus per square foot, per hour.

Insulation: A material with a high resistance (R-value) to heat flow.

Isogonic Chart: A chart that shows magnetic compass deviations from true north.

Isolated Solar Gain: A method of passive solar heating whereby sunlight is captured by a separate, glazed, usually unheated space such as a greenhouse or an atrium. From that space, the heat is distributed to the house. Isolated gain also includes a type of flat-plate collector that uses the natural tendency of heated air to rise as the means of circulating it into the living space. This is called a *thermosiphon* collector.

Langley: A measure of solar radiation, equal to 1 calorie per square centimeter.

Life Cycle Cost Analysis: The measurement and accounting for capital, interest, and operating costs over the useful life of the solar system, compared to the same costs without the solar system.

Microclimate: The variation in regional climate at a specific site caused by topography, vegetation, soil, water conditions, and construction.

Movable Insulation: Insulation that is closed over window areas to heat loss at night or during overcast periods. It may also

be used to reduce heat gain in the summer.
Nocturnal Cooling (Night Sky Radiation): A method of cooling through the radiation of heat from warm surfaces to a clear night sky.
Non-Renewable Energy Source: A mineral energy source that is limited in supply, such as fossil (gas, coal, oil) and nuclear fuels.
Passive Solar Energy Systems and Concepts: Passive solar heating systems, and passive concepts, generally involve energy collection through south-facing glazed areas, energy storage in the mass of the building or in special storage elements, energy distribution by means of such natural processes as convection, conduction, or radiation with minimal use of low-power fans or pumps, and a method of controlling both high and low temperatures, and energy flows. Passive cooling applications usually include methods of shading collector areas from exposure to the summer sun, and provisions to induce ventilation to reduce internal temperatures and humidity.
Payback: The time needed to recover the original investment in a solar energy system through savings in energy costs.
Peak Load: The maximum instantaneous demand for electrical power, which determines the generating capacity required by a public utility.
Percent Possible Sunshine: The amount of radiation available compared with the amount that would be present if there were no cloud cover, usually measured on a monthly basis.
Preheat: The use of solar energy to heat a substance (such as household hot water) part-way to the desired temperature, prior to heating it to a higher temperature by conventional means.
Radiation: The process by which energy flows from one body to another, when the two bodies are separated by a space, even when a vacuum exists between them.
Renewable Energy Source: Solar energy and other energy forms—such as wind, biomass, and hydro—that derive from processes that the sun initiates.
Re-Radiation: Radiation resulting from the emission of previously-absorbed radiation.
Retrofit: To add solar heating to an existing house that pre-

viously used a conventional fuel.
Rock Bed: A heat storage container filled with rocks, pebbles, or crushed stone.
R-Value: The capacity of a material to resist the flow of heat. The term is used to describe the heat-saving value of building insulation and other construction materials.
Solar Access or Solar Rights: The access to sunlight at a given location, usually the south-facing wall or roof of a building, and the ability to protect that access from adverse developments (the construction of high buildings, the planting or growth of shade trees, for example) on land to the south, over which the sun must pass to reach that location.
Solar Collector: A device or building component that collects solar radiation and converts it to heat.
Solar Degradation: The process by which exposure to sunlight causes materials or components to deteriorate.
Solar Fraction: The percentage of a building's heating requirements that can be met by solar energy.
Solar Gain: The absorption of heat from the sun. The amount of solar radiation (in Btus) received on a surface.
Solar Greenhouse: A specially-designed and constructed greenhouse that can be attached to a house to provide heat to the house.
Solar Mass Wall: See *Trombe Wall*.
Solar Radiation: Electromagnetic energy radiated from the sun; visible light and infrared light are the parts of the electromagnetic spectrum used by solar energy systems.
Solarium: A living space enclosed by glass; a greenhouse.
Stack Effect: The draft caused by the rising of hot air over a darkened surface; used to provide summer ventilation in some passive solar houses.
Stagnation: Trapped heat, incapable of movement.
Storage: The component that absorbs collected solar heat and stores it for later use.
Storage Capacity: See *Heat Storage Capacity*.
Sunspace: A living space enclosed by glass; a solarium or greenhouse.
Sun-Tempering: Another passive solar heating method, not

considered a true passive approach because there is no provision for heat storage. For example, a sun-tempered hosue could look much like a direct-gain house with large south-facing windows to collect sunlight, but it would have no massive concrete floors or water-filled containers to absorb and hold heat. This poses two problems, the first being possible overheating during the day, and the second being the lack of stored heat for use at night.

Thermal Chimney: A vertical cavity through which heated air moves as a result of the stack effect. Used as a means of passive solar heat distribution or induced summer ventilation.

Thermal Envelope: The insulated walls, attic, and floors, as well as storm doors and windows, that enclose a building.

Thermal Lag: In an indirect gain system, the time it takes for collected heat to move from the outer collecting surface (of, for example, a Trombe wall) to the inner radiating surface.

Thermal Mass: Building materials like brick, concrete, or adobe, and water in containers, used in passive solar houses to absorb and store heat.

Thermal Radiation: Electromagnetic radiation emitted by a warm body.

Thermal Resistance: See *R-Value*.

Thermosiphoning: Heat transfer through a fluid (such as air or liquid) by means of currents that result from the natural tendency of heavier, cool fluid to fall and the natural tendency of lighter, warm fluid to rise.

Time-Lag Heating: A process by which a building is heated by delayed solar energy, through the use of the heat-absorbing and heat-transfer properties of massive materials.

Trombe Wall: A passive solar heating system in which a masonry wall collects, stores, and distributes heat. Masonry, typically 8 to 16 inches thick, blackened, is exposed to the sun behind glazing.

Trombe Wall Cavity: The space between a solar mass wall and its exterior glazing in which air is heated. This air will rise and may be vented into the building's interior for space heating.

Ultraviolet Radiation: Electromagnetic radiation with wavelengths slightly shorter than visible light.

U-Value: The capability of a substance to transfer heat. U value describes the conductance of a material of composite of materials used in constructing a building.

Vapor Barrier: A waterproof liner used to prevent the passage of moisture through the interior wall of a building so that insulation will not become damp and therefore ineffective.

Ventilation, Induced: The mechanically-assisted movement of air through a building.

Ventilation, Natural: The unassisted movement of fresh air through a building.

Water Wall: A variation on the Trombe wall, in which water filled containers are positioned behind south-facing glazing to absorb and store solar heat.

Zoned Heating: The independent control of temperature for one room or a group of rooms, rather than the whole building.